Zika Virus Disease

From Origin To Outbreak

Zika Virus Disease
From Origin To Outbreak

Edited by

Adnan I. Qureshi

Academic Press is an imprint of Elsevier
125 London Wall, London EC2Y 5AS, United Kingdom
525 B Street, Suite 1800, San Diego, CA 92101-4495, United States
50 Hampshire Street, 5th Floor, Cambridge, MA 02139, United States
The Boulevard, Langford Lane, Kidlington, Oxford OX5 1GB, United Kingdom

Notices
Knowledge and best practice in this field are constantly changing. As new research and experience broaden our understanding, changes in research methods, professional practices, or medical treatment may become necessary.

Practitioners and researchers must always rely on their own experience and knowledge in evaluating and using any information, methods, compounds, or experiments described herein. In using such information or methods they should be mindful of their own safety and the safety of others, including parties for whom they have a professional responsibility.

To the fullest extent of the law, neither the Publisher nor the authors, contributors, or editors, assume any liability for any injury and/or damage to persons or property as a matter of products liability, negligence or otherwise, or from any use or operation of any methods, products, instructions, or ideas contained in the material herein.

Library of Congress Cataloging-in-Publication Data
A catalog record for this book is available from the Library of Congress

British Library Cataloguing-in-Publication Data
A catalogue record for this book is available from the British Library

ISBN : 978-0-12-812365-2

For information on all Academic Press publications
visit our website at https://www.elsevier.com/books-and-journals

Working together
to grow libraries in
developing countries

www.elsevier.com • www.bookaid.org

Publisher: Sara Tenney
Acquisition Editor: Linda Versteeg-buschman
Editorial Project Manager: Joslyn Chaiprasert-Paguio
Production Project Manager: Priya Kumaraguruparan
Cover Designer: Matthew Limbert

Typeset by SPi Global, India

Contents

Contributors

Omar Saeed (Primary Coordinator), Zeenat Qureshi Stroke Institute, St. Cloud, MN, United States

Mohammad R. Afzal Zeenat Qureshi Stroke Institute, St. Cloud, MN, United States

Asad Ahrar Spectrum Health/Michigan State University, Grand Rapids, MI, United States

Morad Chughtai Cleveland Clinic, Cleveland, OH, United States

Ahmed Hassan Zeenat Qureshi Stroke Institute, St. Cloud, MN, United States

Muhammad Fawad Ishfaq Zeenat Qureshi Stroke Institute, St. Cloud, MN, United States

Iryna Lobanova Bogomolets National Medical University, Kiev, Ukraine

Muhammad Ibrahim Malik Upstate University Hospital, Syracuse, NY, United States

Ahmed A. Malik Zeenat Qureshi Stroke Institute, St. Cloud, MN, United States

Mohtashim Arbaab Qureshi Zeenat Qureshi Stroke Institute, St. Cloud, MN; Texas Tech University of Health Sciences Center at El Paso, El Paso, TX, United States

Ihtesham A. Qureshi Texas Tech University of Health Sciences Center at El Paso, El Paso, TX, United States

Mushtaq H. Qureshi Texas Tech University of Health Sciences Center at El Paso, El Paso, TX, United States

Muhammad Akmam Saleem Mercy health System, Janesville, WI, United States

Author Biographies

Omar Saeed graduated from medical school in Pakistan after which he pursued his interest in the field of medicine focusing on neurology and neuroimaging. He was given the opportunity to work with Dr. Adnan Qureshi at the Zeenat Qureshi Stroke Institute where he worked as a clinical research fellow. His main interests were stroke, intracerebral hemorrhage, and neuroimaging, among others. During his time at the stroke institute, Dr. Saeed has coauthored several scientific publications in prestigious journals including *Journal of Neurosurgery*, *Journal of Cerebrovascular Disease*, *Journal of Neuroimaging*, and *Journal of Vascular and Interventional Neurology*. He was also given the opportunity to present at both national and international forums including the International Stroke Conference and the World Masters China Tour and Neuro Interventional International Forum. Furthermore, Dr. Saeed has coauthored numerous research abstracts accepted to both the International Stroke Conference and the American Academy of Neurology annual meeting. His future hopes are to continue focusing on both the clinical and research side of medicine. He served as the primary coordinator for the book on Ebola virus disease that was published in January of 2016 and was also the primary coordinator for the book on Zika virus disease.

Mohammad Rauf Afzal is working as a clinical research fellow at Zeenat Qureshi Stroke Institute. His research interests include stroke epidemiology, intra-arterial and intravenous thrombolytic use in acute ischemic stroke, and neurocritical care. His work has been presented at International Stroke Conference (ISC) and American Academy of Neurology (AAN). He has also coauthored scientific publications in prestigious journals. Along with his research activities, Dr. Afzal has worked as an assistant to core image analyst and as an assistant to primary study coordinator in Hennepin County Medical Center (HCMC) Minneapolis, MN, for Antihypertensive Treatment of Acute Cerebral Hemorrhage (ATACH-II) trial. He continues his research work at Texas Tech University, El Paso, as clinical research fellow.

Ahmed Hassan received his bachelors in medicine and surgery at Cairo University and went on to complete a masters/residency in anesthesiology, critical care, and pain management. During this time, he worked at Cairo University Hospitals "Kasr Al Ainy." He has participated in several studies with emphasis on sepsis in patients because it continues to be a major issue in developing countries like Egypt where he trained. Dr. Hassan was appointed as an assistant lecturer at Cairo University and used to teach medical students, interns, and first-year anesthesiology residents. He joined Dr. Adnan I. Qureshi's research team and is now in the process of completing several research protocol.

Muhammad Akmam Saleem completed his graduate studies at Nishtar Medical College in Multan, Pakistan. He earned his postdoctoral fellowship at the Zeenat Qureshi Stroke Research Institute where he served as the coinvestigator of the prospective clinical trials, analyzed data from national level data sets, and participated in the scientific writing of numerous manuscripts for peer-reviewed journals. He represented his institute at the International Stroke Conferences and the American Academy of Neurology Annual Meetings in the year 2016 and 2017. His contribution to the European Stroke Conference 2016 in Barcelona, Spain, was well received and shared broadly on social media. His work in his 10 research articles in the prestigious journals and 18 published scientific abstracts that address the current issues faced by clinicians. His excursion as a radio presenter has afforded him a unique opportunity to voice his concerns regarding overburdened health care and to advocate for equality in health services. His interest in a holistic approach to patient care bodes well with the biopsychosocial method that stands at the top of population needs. Currently, he is a first-year family practice resident at Mercy Health System in Janesville, Wisconsin, where he looks forward to continuing his productive involvement in evidence-based medicine.

Mohtashim Arbaab Qureshi is a medical doctor by educational qualification and an aspiring neurologist. He is a clinical research fellow under the tutelage of Dr. Adnan I. Qureshi at Zeenat Qureshi Stroke Institute, where he is currently coauthoring *The Atlas of Diagnostic and Interventional Fetal Neurology*. Currently, he is appointed as a research fellow at the Texas Tech University of Health Sciences, El Paso, in the department of neurology, where he is a part of several retrospective and prospective clinical studies. His research interests include stroke and cerebrovascular diseases, interventional neuroradiology and endovascular neurosurgery, and neurocritical care and fetal neurology. He has coauthored several peer-reviewed publications. Dr. Qureshi's work has been presented at several national and international conferences like International Stroke Conference (ISC), American Academy of Neurology (AAN) annual meeting, Society of Vascular and Interventional Neurology (SVIN) meeting, European Stroke Conference (ESC), European Stroke Organization Conference (ESOC), and World Intracranial Hemorrhage Conference (WICH). Besides his research interests, Dr. Qureshi is affiliated with the International Society of Interventional Neurology (ISIN) and International Congress of Interventional Neurology (ICIN) as an associate in the web education and communications committee.

Asad Ahrar graduated from Foundation University Medical College in 2013. After completing his intern year from Fauji Foundation Hospital, he moved to the United States to further his career in medicine. Since then, he has been working as a clinical research fellow at Zeenat Qureshi Stroke Institute. His research interests include stroke epidemiology, stent retrievers in mechanical thrombectomies, intra-arterial and intravenous thrombolytic use in acute ischemic stroke, and neurointervention. He recently describes a new technique that allows visualization of vasa nervorum supplying the nerves through his publication in *Journal of Neuroimaging*. His work has been presented at both the International Stroke Conference and American Academy of Neurology Conferences. He is currently a first-year neurology resident at Spectrum Health/ Michigan State University in Grand Rapids, Michigan.

Morad Chughtai graduated *summa cum laude* from American University of Antigua, College of Medicine in 2014. As a medical student, he served as a teacher's assistant for multiple subjects throughout the duration of his basic science education, including anatomy, physiology, biochemistry, genetics, pathophysiology, and neurosciences. After obtaining his medical degree, he worked under the instruction of Dr. Adnan I. Qureshi as a clinical research fellow for a year's duration. During that time, he traveled to Guinea during the peak of the Ebola virus infection outbreak and coauthored several peer-reviewed publications, one being published in the prestigious journal *Clinical Infectious Diseases*.

Subsequently, he obtained a fellowship at the Rubin Institute of Advanced Orthopedics at Sinai Hospital of Baltimore, in Baltimore, Maryland, in 2015, where he continued his research in the field of orthopedic surgery, under the tutelage of Dr. Michael A. Mont. He subsequently followed Dr. Mont to Cleveland Clinic and continued as a research fellow with him. He now has authored or coauthored over 70 research articles, including one published in the prestigious journal *The Lancet*. In addition, he has authored several textbook chapters in *Orthopaedic Knowledge Update* and *International Neurology*.

Furthermore, his excellence as a teacher's assistant during his time at American University of Antigua, College of Medicine has allowed him to obtain a staff position there as a visiting professor in the department of anatomy, where he regularly teaches system-based anatomy to first- and second-year medical students. He hopes to carry his passion of teaching and research throughout his medical career and eventually practice in an academic setting. He is currently a first-year orthopedic surgery resident at Cleveland Clinic, in Cleveland, Ohio, where he continues his passion in research while honing his clinical acumen.

Muhammad Ibrahim Malik attained his medical degree from Batterjee Medical College in Jeddah, Saudi Arabia. He completed electives in cardiothoracic surgery and surgical ICU at Cleveland Clinic, in Cleveland, Ohio. After completing one year as an intern doctor in Al-Noor Specialist Hospital in Makkah, Saudi Arabia, he moved to the United States to get his American medical license. During this time, Dr. Malik volunteered at the Rahma Free Medical Clinic in Syracuse, NY, working with doctors to provide care to patients from low-income neighborhoods. Dr. Malik is currently a research assistant in the department of neurology at Upstate University Hospital, in Syracuse, NY, and he is working on completing his US medical licensing exams.

Ihtesham A. Qureshi is a clinical research fellow in neurology at Zeenat Qureshi Stroke Institute. His main areas of research include stroke and cerebrovascular diseases, neurointervention, and infectious disease. He has coauthored numerous scientific research papers published in renowned medical journals and has also presented scientific research abstracts at various reputed national and international medical conferences.

He was also a former medical field doctor at Doctors without Borders/ Médecins Sans Frontèires (MSF) and was involved in operational research on tuberculosis and malaria. He was the principal investigator for the first study ever done to assess memory decline among Ebola virus disease survivors using mini-mental scale examination. Currently, Dr. Ihtesham Qureshi is working as a resident physician in neurology at Texas Tech University of Health Sciences Center at El Paso, Texas.

Iryna Lobanova, MD (2006) and PhD (2011). Assistant professor of the neurology department of O.O. Bogomolets National Medical University, Kiev, Ukraine (since 2011), from which she graduated with honors in 2006. The main research areas are vascular and demyelinating diseases of the nervous system. She is an author of more than 78 scientific works and owner of five patents.

Ahmed A. Malik is a medical doctor by education and is currently a clinical research fellow at the Zeenat Qureshi Stroke Institute. After finishing some undergraduate coursework at SUNY Stony Brook University, Dr. Malik, guided by a desire to understand health care in the developing world, went to the Shifa College of Medicine in Islamabad, Pakistan, to complete his medical education. Upon his return to the United States, he worked on a volunteer basis with doctors at private clinics before joining Zeenat Qureshi Stroke Institute. As a clinical research fellow at Zeenat Qureshi Stroke Institute, Dr. Malik has worked under the mentorship of Dr. Adnan I. Qureshi to author numerous scientific research papers published in renowned medical journals. Dr. Malik has also presented scientific research papers at national and international medical conferences and is a reviewer for the *World Journal of Pediatrics* and a contributor to the *Journal of Vascular and Interventional Neurology*. In his free time, he likes to write and is currently writing a few works of fiction for publication. Dr. Malik also recently completed his masters in public health form SUNY Upstate and Syracuse University.

Muhammad Fawad Ishfaq, MD, obtained a Bachelor of Medicine, Bachelor of Surgery (MBBS) degree from National University of Science and Technology (NUST). Dr. Ishfaq is currently working as a clinical research fellow at Zeenat Qureshi Stroke Institute and also a resident physician at University of Tennessee Health Science Center. His research interests include various management protocols in ischemic and hemorrhagic stroke. His work has been frequently presented at International Stroke Conference (ISC), American Academy of Neurology (AAN) annual meeting, and European Stroke Organization (ESO) conference. He has also coauthored scientific publications in many prestigious journals.

Mushtaq H. Qureshi is working as a senior clinical research fellow at Zeenat Qureshi Stroke Institute. He has authored and coauthored several scientific publications which are published in various prestigious journals. He has also made several platform and poster publications in various national and international meetings. He is also the managing editor of *Journal of Vascular and Interventional Neurology* and also serves his role as the head of imaging department in a large phase III clinical trial, which is funded by National Institutes of Health. Currently, Dr. Qureshi is a neurology resident at Texas Tech University in El Paso where he continues his clinical and research work.

Preface

The book "*Zika virus disease: from origin to outbreak*" is another exciting contemporary work after our previous book "*Ebola virus disease: from origin to outbreak*" which was very well received. However, the Zika virus infection outbreak was completely different from the previous Ebola virus disease outbreak. Zika virus disease is a mild disease with most infected patients experiencing no or nonspecific symptoms with complete recovery. Dying from the disease is almost unheard of. And yet, World Health Organization declared Zika virus disease a "Public Health Emergency of International Concern." As the readers will learn through well chronicled progression of events over half a century in our book, the mild self-limiting nature of Zika virus infection made it difficult to identify the disease among population and harder to justify allocating resources to control the global spread. The world started expressing prominent concerns when Zika viral infection during pregnancy was linked to neurological abnormalities occurring in fetuses. Our book will lead the readers through comparisons with previous and related viral infections so the unique aspects of this outbreak are highlighted. The new face of a viral outbreak is not going to be characterized by isolation camps populated by dying patients and medical professionals with daunting protective gear. The invisible face of the outbreak will be concerned potential parents living through the threat of having a new born with serious lifelong disability with no means to prevent this occurrence (unlike other infections in pregnancy). Our book will provide an up to date description on Zika viral infections, pathophysiology, and therapeutics. The readers will appreciate the social, financial, and psychological aspects of such a disease through well-articulated data presented in our book. The reader may be left to wonder how many other viruses are out there that we have ignored as insignificant and thus allowed them to progress to a full blown outbreak like Zika virus infection.

Introduction

As the havoc wrought by Ebola virus disease tapered off, widespread anxiety about a global outbreak began to calm down. It did not appear that another outbreak or epidemic would be seen in the future, at least not in the near future.

In the latter half of 2015, however, South America reported an outbreak of a mosquito-borne virus known as "Zika virus infection." Simultaneously, an unusual increase in the number of newborns with microcephaly (small head and brain) was noticed and linked to the epidemic of Zika virus infection of pregnant women. Shortly thereafter, reports confirmed that Zika virus infection was indeed the reason for the microcephaly in newborns. By November 2015, Brazil had declared the epidemic a national public health emergency, and the World Health Organization (WHO) announced that there was a high potential for the virus to spread further in the Americas. The concerns were brought forward with unprecedented urgency when 2016 Olympics were hosted in Rio de Janeiro, Brazil, striking fear in the hearts of health-care professionals, athletes, and attendees who were concerned about becoming infected or, more alarmingly, bring the virus back to their parent countries such as North America. The most concerning aspect of the outbreak was the potential to increase the proportion of disabled children despite a high survival from the infection.

This book will provide and discuss several aspects of the Zika virus infection, including its history, how the virus is acquired and spread, viral structure with relevance to virulence, associated symptoms, post-infectious consequences, and treatment options. In addition, the health-care perspective (both globally and within the United States), economic and political aspects, and psychosocial implication of Zika virus infection will be discussed. We hope this book will provide a concise narration about various aspects of this outbreak.

Chapter 1

Origin of Zika Virus Disease

The rapidly evolving outbreak of Zika warns us that an old disease that slumbered for 6 decades in Africa and Asia can suddenly wake up on a new continent to cause a global health emergency.
WHO Director-General, Dr. Margaret Chan, May 23, 2016.

It has long been agreed upon that mankind will always fear what it is unable to understand. Many of these fears stem from the thought of bodily harm that might lead to death or harm to future offspring. Most recently, the growing fear of an old infection long thought to be dormant has caused havoc and mayhem in a new subset community, most notably the pregnant women.

Zika virus disease has been present for decades and was first discovered more than half a century ago. But where did it come from and how did it first manifest in those affected? These are the questions we will try to answer as we try to understand the origins of the Zika virus disease.

Zika virus is a type of virus that belongs to the Flavivirus family, which also includes viruses such as the West Nile virus, Dengue virus, tick-borne encephalitis virus, and Yellow Fever virus. Zika virus disease is a mosquito-borne disease, meaning it spreads most notably by the bite of an infected mosquito. These mosquitos can bite at anytime during the day or night. Currently, there are no vaccines available for the disease.[1] The presence of this disease outside of the United States has been documented for decades. But panic spread when potentially infected cases inside the US territory were identified, especially in the

Zika Virus Disease. https://doi.org/10.1016/B978-0-12-812365-2.00002-0

1

southernmost states. This disease has had numerous outbreaks spanning from 1947 to the most recent outbreak in 2015, which began in Brazil and has spread to the US state of Florida and others.[2]

AN ACCIDENTAL OR INCIDENTAL FINDING?

First Encounter

During the early 1940, there were many viruses being discovered in the mosquito population. Scientists from the National Institute for Medical Research in London, George W. A. Dick, and from the Division of Medicine and Public Health and the Rockefeller Foundation in New York, S. F. Kitchen and Alexander Haddow are considered to be the discoverers of the Zika virus disease.[3] It can be argued that it may have been an accidental finding rather than a specific attempt at isolating the virus itself. Admittedly, many major discoveries throughout the centuries have been accidental, like Viagra or penicillin. The scientists were originally in Uganda to study and isolate Yellow Fever virus samples. According to existing sources, these two scientists isolated the Zika virus on two separate occasions approximately 9 months apart by two different methods.

Zika Forest is an isolated tropical forest approximately 4 miles from a centrally located town in Uganda called Entebbe. The forest is considered a property of Uganda Virus Research Institute, Entebbe (UVRIE), and is thus restricted from use by any foreign scientists. The forest is about 62 acres and is bordered by a swamp that is part of Lake Victoria—the lake itself is roughly 500 yards wide and 1 mile in length (Fig. 1.1).[3,4] Due to the high levels of immunity to Yellow Fever in monkeys of the Entebbe peninsula, the scientists had chosen Zika Forest

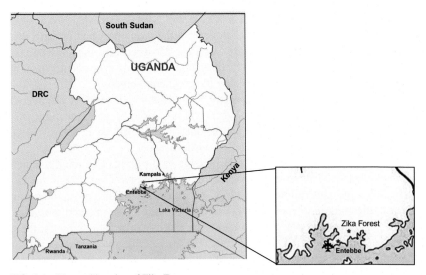

FIG. 1.1   Map and location of Zika Forest.

as a study location. Alexander Haddow, a virologist, had set up the sentinel rhesus monkey program where rhesus monkey were caged high in the canopy.[5] Supposedly these treetop locations, coupled with the temperate environment, were an ideal environment for virus isolation or transmission by mosquitos.

The scientists took temperature of six sentinel platform locations during April 1947. On April 18, 1947, one of the Rhesus monkeys, specifically Rhesus 766, exhibited a temperature of 39.7°C on the subsequent day, April 19, a second reading of the same Rhesus monkey (Rhesus 766) was 40°C.[3] Although the Rhesus 766 failed to show any further signs of the disease besides the elevated temperature, blood samples were taken and serum was injected into other mice (intraperitoneally and intracerebrally) and Rhesus 771 (subcutaneously) for further investigation. Those mice who were inoculated intracerebrally showed signs of sickness, but the Rhesus 771 and intraperitoneally inoculated monkey and mice did not manifest any symptoms.[3] This was to be later documented as potentially the first ever Zika virus disease manifestation in an animal.

Second Encounter

The second specimen of Zika virus was to be isolated from the actual mosquito itself. In January 1948, about 10 months after the first possible detection of this new type of virus, the same treetop platform technique was used to catch mosquitos. These mosquitos were then brought to the laboratory in Entebbe where they underwent further testing for species or genera identification. The most prominent of the mosquito species *Aedes africanus* (Fig. 1.2) was isolated

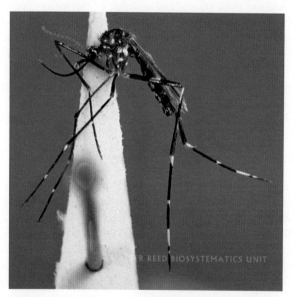

FIG. 1.2   *Aedes africanus* (Ref: http://wrbu.org/mqID/mq_medspc/AD/AEafr_hab.html).

separately due to its previous implication in Yellow Fever transmission. The methods by which they were prepared for further testing have been outlined by Dick et al.[3] Two different types of inoculates were produced and one of them, called E/1/48 lot made by using 86 *A. africanus* species of mosquito, was an actual strain of virus responsible for Zika virus disease. These supernates would later be injected into mice and other rhesus monkeys although only mice demonstrated signs of sickness with elevated temperatures. Interestingly, when serum taken on the 8th day from infected monkeys (who did not show signs of disease) was injected into mice, the mice ended up dying, but when serum taken on the 10th day was injected, all the mice remained alive.[3] It was later found that these Rhesus monkeys were infected with Zika virus by the mosquito's inoculants and thus became the second isolation of this previously unknown virus.

All these findings were later cross-tested with other potential viruses for serological specificity. To make sure that this strain of virus was indeed something new, scientists in Uganda ran cross neutralization tests with Yellow Fever, Dengue virus, Theiler's encephalomyelitis virus, Mango encepholmyleitis, and Uganda S virus. Additionally neutralization tests for Easter equine, Western equine, and St. Louis and Japanese B encephalitis viruses were performed in New York, but none demonstrated any neutralizing effect. This was the first ever discovery of what is now known as the Zika virus.[3] The findings in mosquitos affirmed the notion that this virus is, for the most part, an arbovirus (an arthropod-borne virus).

Researchers concluded that Zika virus was unequivocally found in the Zika Forest in both Rhesus monkey and mosquitos. The time of infection as suggested was between April 1947 and January 1948 meaning the infected arthropods carried the virus throughout and that of the many species of "biting insects" studied *A. africanus* was the only one infected with the Zika virus. The only question to answer at this time was whether the mosquitos were actual vectors of the disease, because *A. africanus* was a known vector of Yellow Fever virus at that time.[6] It was also not clear whether the Zika virus had infected and/or manifested disease in humans, especially with the absence of antibodies in local Zika Forest residents. This lack of immune response may have been due to the fact that local residents did not have much contact with the forest.

THE FIRST HUMAN ZIKA VIRUS ISOLATIONS

Although outbreak of Zika virus had not yet occurred, the first ever evidence that pointed toward the presence of this virus in a human being was described in 1952. In this study, eight viruses were isolated to be studied in human sera. Included in them were Mengo virus, Bwamba fever, Nataya, West Nile, and Zika virus. Out of the 297 residents of Uganda and Tanganyika, Bwamba, Zika, Ntaya, and Uganda S viruses were the most prevalent with adequate immunity and antibodies against those viruses. Zika virus antibodies were neutralized in approximately 38 of the 297 specimens tested.[7] Further investigation demonstrated the wide range of geographical distribution of these viruses including Zika virus which meant that it was not just isolated to the Zika Forest near Entebbe.

THE FIRST CLINICAL OCCURRENCE

The isolation of Zika virus from the serum of the above-mentioned population was also incidental and cannot be considered a true outbreak, occurrence, or manifestation of the disease itself. The actual first case of Zika virus with probable disease manifestation was originally described in 1954. It was described by MacNamara during an investigation of a jaundice outbreak believed to be caused by Yellow Fever in Eastern Nigeria.[8] During this outbreak there were three patients (Zika virus isolation from one patient and high serum antibody titer in the other two patients) in whom Zika virus was the presumed infectious agent. All patients were observed in outpatient setting so follow-up was not possible at that time. Of these three cases the first two were Nigerian men aged 30- and 24-years old and the third case in which Zika virus was isolated from was a 10-year-old African girl. The first case, who was 30-years old man, was ill for 14 days and presented with cough, joint pain, and fever. He also had mild jaundice but all other findings were normal with no malarial parasite detected in blood. The second case was the 24-year-old man, who also complained of similar symptoms of fever, headache, and joint pain. He also had pain behind his eyes and loose, watery stool although this patient was not jaundiced. The third and final case was a 10-year-old girl who presented with fever and headaches but did not show signs of jaundice—her blood did, however, contain numerous malarial parasites.[8] This could potentially be the first description of how Zika virus manifests in human beings (i.e., fever, headache, joint pains, potentially jaundice, and loose stools).

Despite having studied patients with definitive Zika virus disease, the clinical picture presented by this infection was unclear. It was not scientifically certain whether the African girl and two other patients who showed signs of rising serum antibodies to the virus were actually having Zika virus disease manifestations. It is possible that the patient who did show signs of liver dysfunction manifesting as jaundice was not actually Zika virus infection but a concomitant infection with a different virus. But, even mouse models had suggested jaundice can occur with Zika virus infections.

To provide a clinical picture of this virus, a volunteer was selected to be infected with the virus in 1956. A 34-year-old European man who was residing in Nigeria for a few months volunteered to be infected with the Zika virus. He was then be closely monitored from the day of inoculation at regular intervals. After about 82 h of inoculation the patient began to experience generalized headaches, fever, and malaise. This continued for 2 days after which he experiences a sharp rise in temperature to 100.5 F and the headache and malaise increased in severity with accompanying nausea and vertigo. On day 7, the patient was feeling quite well; the only treatment that was given to him was aspirin for symptom control.[9] This was also one of the first times that *Aedes aegypti* (Fig. 1.3) mosquitos

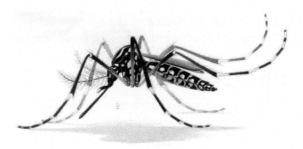

FIG. 1.3 *Aedes aegypti* mosquito.

were used instead of the *Aedes africanus* which were originally thought to be main vectors of the Zika virus disease proving that this virus was capable of spreading through different mosquito species.[9,10]

BETWEEN THE 1960S AND 2000

Equatorial Africa

From the start of the Cuban Missile crisis to the end of Presidents Bill Clinton's impeachment trials, and while the world was distracted by major wars, Equatorial Africa was being silently overwhelmed by the Zika virus disease. During this time, there had been numerous encounters noted by scientist of this virus in mosquitos, monkeys, and humans. The virus was sporadically isolated in numerous African countries, which demonstrated the vast geographical dissemination of this virus.[11–14] During these isolations, patients rarely if ever demonstrated any signs of the disease, and the virus was merely isolated from blood samples.

In 1964, Zika virus infection did manifest in a 28-year-old European man who had been residing in Uganda for approximately 2 months. This patient had been vaccinated against Yellow Fever virus at the time but developed typical symptoms of Zika virus infection.[8,15] The patient developed a frontal headache, malaise, fever, and general body aches during the infection. But he also developed a maculopapular rash that spread all throughout his body involving the soles of his feet and palms of his hands.[15] This presentation could also have been a misdiagnosis or co-infection with another virus at time.

Later, between 1964 and 1970, arthropod-borne viral infections were being studied in Nigeria where scientist found isolation of numerous viruses in humans. The majority of these were isolated during the wet or rainy season when mosquitos have better access to water for egg laying and reproduction.[14] Although Zika virus was a less frequent isolation, it was present none the less in several specimens.

Between 1970 and 1975, a special Zika virus study was carried out in the Oyo State, which is located in southwestern Nigeria. During this study, two Zika virus isolations were made from 10,778 blood specimens. Both of these

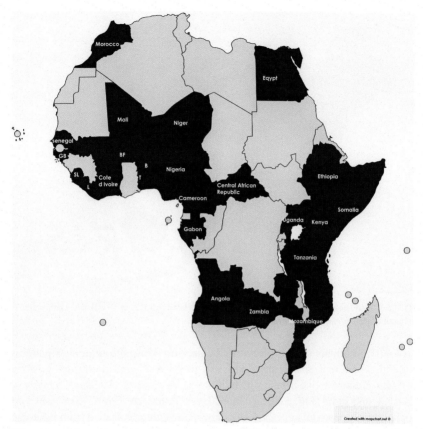

FIG. 1.4   African countries where Zika virus disease has been reported up to 2016. *GB*, Gabon; *SL*, Sierra Leone; *L*, Liberia; *BF*, Burkina Faso; *B*, Benin; *T*, Togo.

isolation's were derived from young children aged 2- and 10-years old and both isolations were during rainy seasons. These patients demonstrated the classic symptoms of febrile illness, headache, and malaise.[16] The most interesting part of this study was the amount of neutralizing antibodies found in the Nigerian communities. It showed that 40% of Nigerians had Zika virus neutralizing antibodies, which could explain why the virus is present all over Africa, but clinical illnesses were extremely rare (Fig. 1.4).

ZIKA VIRUS DISEASE OUTSIDE OF EQUATORIAL AFRICA

Asia

Africa seems to be designated as the birth of many viruses—Ebola, Yellow Fever, Chikungunya, Zika, and other viruses. But like all these viruses Zika virus did not remain endemic to Africa. Just like the recent Ebola virus disease outbreak during

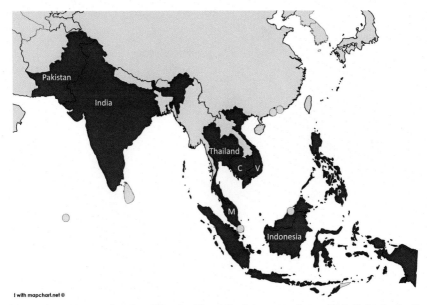

I with mapchart.net ©

FIG. 1.5   Asian countries where Zika virus disease has been reported up to 2016. *C*, Cambodia; *V*, Vietnam; *M*, Malaysia; *P*, Philippines.

2014–2015, humans spread the virus themselves. Mosquito migration patterns could also be a reason why the Zika virus was able to spread outside of Africa to Asian countries like Malaysia, Indonesia, Philippines, Thailand, Maldives, Pakistan, and India (Fig. 1.5).[17–23] In Malaysia in the late 1960s, Zika virus was isolated from approximately 29 *Aedes aegypti* mosquitos collected from Bentong, Malaysia; manifestation of disease was not observed in any individual.[17] In the late 1970s, Flavivirus infected individuals were tested for presence of Zika virus and showed positive titers. These patients demonstrated symptoms of malaise, fever, stomach ache, and anorexia.[18] In the1950s Indian scientists wanted to perform a survey of immunity to a certain type of virus but extended their study to include numerous viruses not usually found in the region. They discovered that approximately 33 of the 196 sera tested positive for Zika virus antibodies, which was one of the first times exposure to virus was demonstrated in India or outside of Africa in general.[21] In the early 1980s, even Pakistan demonstrated positive serum antibodies to Zika virus with an overall prevalence rate of 2.4%.[20]

Pacific Islands

More recently, besides the major outbreak in Brazil, Zika virus has been isolated in numerous other post 21st century countries, including islands in the Pacific such as the French Polynesian Islands.[24] One of the first largest outbreaks of Zika virus infections was on Yap Island, which is part of the Federation State of Micronesia in 2001. In this outbreak, physicians characterized the symptoms of patients as combination of rash, conjunctivitis, and join pains. They identified 49 confirmed cases and 59 probable cases of Zika virus disease in the tested population. Household surveys were later done to detect the extent of infection and the results were staggering. Among the 173 randomly selected houses and 557 household residents tested, 74% demonstrated antibodies against Zika virus suggesting that these individuals had been exposed or had contracted the disease during their lifetime.[25] Not all patients who are exposed became symptomatic. But this remained to be one of the first largest outbreaks of Zika virus disease in recent history. Questions remained as to why an outbreak of such magnitude occurred, even though there had been numerous individuals with confirmed Zika virus exposure and some who even presented with symptoms. This could be attributed to relative immunity built up by constant exposure in the African population and lack of such immunity in the Western Pacific Micronesian people.

The next largest outbreak was so significant that World Health Organization (WHO) stepped in because of the signs of Gullian-Barre syndrome in virus infected patients and viral effects on pregnant women and newborns.[26] In October of 2013, three members of a household in French Polynesia experienced what was then suspected to be a Dengue-like illness with fever, physical weakness, joint pain, rash, and headaches. Two of the three patients also experienced conjunctivitis, a symptom specific to Zika virus infection. During this time the Department of Health at French Polynesia recorded high numbers of primary care visits from patients with these symptoms. After assessing all networked hospitals and clinics, an estimated 19,000 cases of Zika virus disease were evaluated.[27]

In 2014, the health department in French Polynesia began to monitor the spread of this new virus infection in their region. They took samples of patients who presented with symptoms similar to those who were previously diagnosed in October. They also observed patients and took surveys and questionnaires about their symptomatology. In the 6 months of surveillance, authorities found a total of 8750 suspected cases of Zika virus infection throughout the Pacific French Polynesian island territory with a total affected number estimated at 32,000, which represented 11.5% of the population.[28] During this outbreak was also the first time that the possible sexual transmission of Zika virus was discovered.

In December of 2013, 44-year-old man in Tahiti had symptoms of Zika virus disease with malaise, headache, fever, and joint pains. After his symptoms subsided, he had a recurrence of Zika virus infection. He then completely recovered, but later noticed some blood in his semen. Since he had previous history of Zika virus infection, a real-time reverse transcriptase test on his semen was positive for Zika virus.[29] This would be the first confirmed detection of Zika virus in semen with the potential for sexual transmission. In 2008, a man who was infected with the virus in Senegal returned to the United States. He had sexual intercourse with his wife who subsequently developed Zika virus disease symptoms, was confirmed by serological test. His semen, however, was never tested for the presence of Zika virus itself.[30]

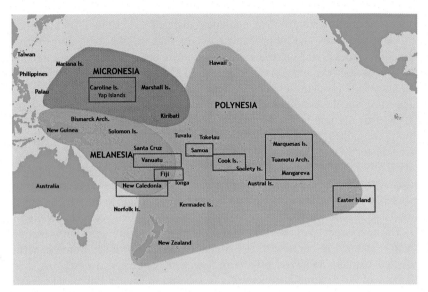

FIG. 1.6   Pacific Island countries where Zika virus disease has been reported up to 2016. *Source: https://commons.wikimedia.org/wiki/File:Pacific_Culture_Areas.jpg.*

The Zika virus disease had also spread to neighboring islands including Cook Island, New Caledonia, and Easter Island, who reported confirmed cases of the virus infection (Fig. 1.6).[24,31,32] In new Caledonia, the only previous vector-born virus infection ever reported was Dengue fever virus, which has been prevalent in the since 2001.[33] The first case of Zika virus disease was reported in November of 2013 in New Caldeonia. The disease was suspected to have been imported from French Polynesia with the first autochthonous case reported by January 2014. In February of 2014, the Health Authorities in New Caledonia officially announced an outbreak. Since then, there have been roughly 14,000 Zika virus disease cases reported in the region confirmed by laboratory testing.

In March of 2014, the Cook Islands declared an outbreak of Zika virus disease with 18 laboratory confirmed cases. This outbreak however was smaller than the one that occurred in other islands with only 900 reported cases, according to WHO in their surveillance report.[34] Various geographical factors could have contributed to the cause of rapid spread of the virus infection in the Pacific islands. The extent of disease's burden may be explained by the climate, patient population, or even difference in vectors present on the islands. Nevertheless one thing was clear, this virus infection was spreading fast and affecting larger populations as the virus continued its migration.

During these outbreaks in French Polynesia, the first possible evidence of Zika virus disease being vertically transmitted was also reported—one in December of 2013 and the other in February of 2014. Both women had symptoms of maculopapular rash, joint pains, and fever, although only one of the newborns displayed signs of disease. Both women and newborns sera were positive for Zika virus disease when serum was tested using Reverse transcription polymerase chain reaction.[35]

MOST RECENT OUTBREAK

South America

South America is a continent located relatively equal on the geographical co-ordinates of Africa. Bordered both by the Pacific and the Atlantic Ocean, it is home to 12 sovereign states, with the largest city of São Paulo, Brazil, having a population of approximately 12 million inhabitants. Zika virus infection is considered to be new to South America. In early 2015, there were reports of patients presenting with a "Dengue-like" viral disease in the eastern part of Brazil mainly in states of Rio Grande do Norte, Bahia, Sergipe, and Paraiba.[36–38] These patients presented with typical symptoms of rash, low fever, headache, joint pains, and edema. Later in March of 2015, the Molecular Virology Laboratory of Carlos Chagas Institute, Oswaldo Cruz Institute, State of Parana, Brazil, received 21 specimens from patients exhibiting those symptoms. These specimens were first tested against Dengue and Chikungunya virus and came out negative. The investigators then tested 2 out of the 21 specimens against Zika virus, specifically the Asian lineage, and the tests were positive. This was to be the first confirmation of Zika virus infection in Brazil. The most common reported symptom in these patients was maculopapular rash, fever, and joint pains.[36]

In the same month, serum samples from 24 patients were obtained from Santa Helena Hospital in Camacari who were given a possible diagnosis of Dengue fever and were subsequently treated for suspected Dengue virus in the emergency department. Serum samples from these patients were later analyzed at the Federal University of Bahia where seven patient samples (29.2%) were positive for Zika virus disease.[37] Surprisingly in both Rio Grande Do Norte and Bahia Brazil, a majority of infected patients were women between ages of 28 and 39 years. As of October 2015, 14 states had confirmed Zika virus infection including Alagoas, Bhia, Ceará, Maranh ã o, Mato Grosso, Pará, Paraíba, Parana, Pernambuco, Piauí, Rio de Janeiro, Rio Grande do Norte, Roraima, and S ã o Paulo. Health Ministires of Colombia also reported its first case in the same month.[39] Since the beginning of the outbreak from December of 2014 to January 2016, Zika virus disease had spread to every region of Brazil including North, Northeast, Central-west, Southeast, and South.[38,40,41] It was estimated by January of 2016 that 440,000 to 1,300,000 suspected cases of Zika virus disease had occurred in Brazil just in 2015.[42]

So how did Zika virus get to Brazil and South American Countries? It was initially assumed that Zika virus was introduced to Brazil during the world cup soccer tournament in 2014 with so many countries competing and visitors flying in to watch and participate. However, since none of the Zika endemic Pacific countries participated, this event could not have caused the spread. In August of 2014, however, the Va'a World Sprints Championship Canoe race was held in Rio De Janeiro, Brazil, where four Pacific countries participated in the competition (French Polynesia, New Caledonia, Cook island, and Easter, Island)—all of these countries had a Zika virus disease outbreak in 2014.[43] This could have been the potential route that caused the Zika virus to enter and spread in South American countries. Of course, the growing fear was how far north this virus infection would spread. Since it was clear that spread of this virus is not just vector borne and that it could be spread by sexual contact, the big fear that remained in Brazil was around Rio de Janeiro hosting the 2016 summer Olympics. Mass gatherings could cause both dissemination of the virus, potential international spread, and affect many Olympic athletes all of whom needed to be strictly monitored and controlled.[44]

ZIKA VIRUS DISEASE AND PREGNANCY

During this outbreak, physicians noticed an increasing number of newborn babies with microcephaly (a rare neurological condition where infant brain does not develop properly and results in smaller than normal head).[45–47] In September and October of 2015, health authorities confirmed an increasing number of infants born with microcephaly in northeast Brazil.[46] On December 15, the Health ministry of Brazil confirmed 134 cases of microcephaly believed to be associated with prenatal Zika virus infection. There were an additional 2165 cases in 549 counties in 20 states that were under investigation.[45] This was confirmed by the Pan American Health Organization (PAHO) when they identified Zika virus RNA by RT-PCR in amniotic fluid samples taken from two pregnant women whose fetuses had ultrasound confirmed microcephaly.[46,47] This discovery caused mass panic in the pregnant population, especially ones in the endemic area.

The Centers for Disease Control and Prevention (CDC) published numerous guidelines that have been continuously updated ever since to help protect all pregnant women.[48–51] All women who are planning to travel to endemic areas were warned about the potential for Zika virus infection and its effect on the fetuses. There were various different checklist for physicians managing pregnancies and patients who were pregnant and had traveled to area with Zika virus infection, pregnant women living in an area with Zika virus infection, and even Zika virus testing for any pregnant women not living in area with the virus infection (Figs. 1.7-1.9).[52] By February 1, 2016, WHO had declared this link between microcephaly and Zika virus was identified as a Public Health Emergency of International Concern (PHEIC). The WHO committee also made numerous recommendations, including surveillance for microcephaly in Zika virus infected areas, long-term measures focusing on research, travel measures such as restrictions to endemic areas, and data sharing of clinical, virology, and epidemiological importance.[53] As of January 5, 2017, according to the WHO situation report, there have been approximately 2564 cases of microcephaly and/or central nervous system malformation due to Zika virus infection or potentially associated with Zika virus infection in 29 different countries.[54] Out of all these cases, more than 89% occurred in Brazil alone.

ZIKA VIRUS INFECTION AND GUILLAIN-BARRE SYNDROME

Another threat that had started to emerge during these Zika virus infection outbreaks were the increasing number of patients who were presenting with

CDC's Response to **Zika**

ZIKA VIRUS TESTING FOR ANY PREGNANT WOMAN NOT LIVING IN AN AREA WITH ZIKA

CDC understands that a pregnant woman may be worried and have questions about Zika virus infection (Zika) during pregnancy. Learn more about Zika virus testing for a pregnant woman and what you might expect if you have Zika during your pregnancy.

How Zika spreads.

A pregnant woman who does not live in an area with Zika can catch the virus from a mosquito bite while visiting an area where mosquitoes are spreading Zika. She can also get Zika through sex with an infected partner. For more information on transmission of Zika, visit www.cdc.gov/zika/transmission.

What CDC knows about Zika virus and pregnancy.

- Zika virus can spread from mother to fetus during pregnancy and around the time of birth.
- Zika virus can cause birth defects and has been linked with other problems in infants.

What CDC doesn't yet know about Zika virus and pregnancy and is researching quickly to find out.

If a woman is infected during pregnancy, we don't know

- How likely it is that the virus will affect her or her pregnancy.
- How likely it is that the virus will be passed to the fetus.
- How likely it is that the fetus, if infected, will have birth defects.
- When in pregnancy the infection might harm the fetus.

www.cdc.gov/zika

U.S. Department of
Health and Human Services
Centers for Disease
Control and Prevention

CS268732A   August 10, 2016

FIG. 1.7 CDC recommendations for pregnant women not residing in Zika virus infection endemic area.

Guillain-Barre syndrome. This is a neurological disease that presents as progressive symmetrical weakness of the limbs with loss of reflexes and symptoms peaks usually within 2-4 weeks.[55] Although the most common causes of this disease are gastrointestinal infection or respiratory infection, it has been linked to other viral hemorrhagic fevers such as Dengue and Chikungunya.[57,58]

It was proposed during the 2013–2014 French Polynesian outbreak that Zika virus could potentially cause Guillain-Barre syndrome.[26] It was later confirmed

CDC's Response to **Zika**

DOCTOR'S VISIT CHECKLIST:

For Pregnant Women Who Traveled to an Area with Zika*

If you are pregnant and have traveled to an area with Zika during your pregnancy or up to 8 weeks before becoming pregnant, you should talk to your healthcare provider, even if you don't feel sick.

Bring this checklist to your visit to make sure you don't forget to discuss anything important.

Here are some topics and questions you may want to discuss with your healthcare provider:

INFORMATION TO SHARE:	QUESTIONS TO ASK:

✓ When did you travel to an area with Zika?
  • Where did you travel?
  • How long did you stay?
✓ In what trimester was your pregnancy when you traveled to an area with Zika?
✓ Did you have any symptoms of Zika during your trip or within 2 weeks of returning?
  • The most common symptoms of Zika are fever, rash, joint pain, and red eyes.
✓ Did your partner travel to an area with Zika?
  • When and where did your partner travel?
  • Did your partner have any signs or symptoms of Zika (including fever, rash, joint pain, or red eyes) when they were on the trip, or after returning?
	✓ Should you be tested for Zika virus?
	  • Pregnant women with possible exposure to Zika virus should be tested for Zika infection, whether or not they have symptoms.
	✓ Do you need an ultrasound?
	✓ Do you need to be referred to a maternal-fetal medicine specialist or a high-risk obstetrics specialist?
	✓ How can you prevent sexual transmission of Zika virus?
	Be sure to ask any other questions or mention concerns you may have about Zika and your pregnancy.

*Check **wwwnc.cdc.gov/travel/notices** for the most up-to-date travel recommendations.

Resource List:
Areas with Zika Virus: www.cdc.gov/zika/geo/
Facts About Microcephaly: www.cdc.gov/ncbddd/birthdefects/microcephaly.html
Zika Virus and Pregnancy: www.cdc.gov/zika/pregnancy/index.html
Pregnant Women: How to Protect Yourself: www.cdc.gov/zika/pregnancy/protect-yourself.html
Mother-To-Baby Website: www.mothertobaby.org/
Zika Virus Prevention: www.cdc.gov/zika/prevention/index.html
Zika and Sexual Transmission: www.cdc.gov/zika/transmission/sexual-transmission.html

U.S. Department of Health and Human Services
Centers for Disease Control and Prevention

CS264360A   October 14, 2016

FIG. 1.8   CDC recommendations for women who have traveled to Zika virus infection endemic area.

that indeed there was a link between the virus and the syndrome. During the French Polynesian outbreak, there were approximately 42 patients at the Centre Hospitalier de Polynésie Française in Tahiti, with Gullian-Barre syndrome associated with Zika virus infection. The current South American countries affected are expected to prepare prepare for an increased number of patients presenting with this syndrome.[59]

CDC's Response to **Zika**

ZIKA VIRUS TESTING FOR PREGNANT WOMEN LIVING IN AN AREA WITH ZIKA

CDC understands that a pregnant woman may be worried and have questions about Zika virus infection (Zika) during pregnancy. Learn more about Zika virus testing for pregnant women and what you might expect if you have Zika virus during pregnancy.

How Zika spreads.

A woman who lives in an area with Zika can get the virus from the bite of an infected mosquito. She can also get Zika through sex with an infected partner. For more information on transmission of Zika, visit www.cdc.gov/zika/transmission.

What CDC knows about Zika virus and pregnancy.

- Zika can spread from mother to fetus during pregnancy and around the time of birth.
- Zika can cause birth defects and has been linked with other problems in infants.

What CDC doesn't yet know about Zika virus and pregnancy and is researching quickly to find out.

If a woman is infected during pregnancy, we don't know

- How likely it is that the virus will affect her or her pregnancy.
- How likely it is that the virus will be passed to the fetus.
- How likely it is that the fetus, if infected, will have birth defects.
- When in pregnancy the infection might harm the fetus.

www.cdc.gov/zika

U.S. Department of Health and Human Services
Centers for Disease Control and Prevention

CS269732A   August 11, 2016

FIG. 1.9   CDC recommendation for women living in Zika virus infection endemic area.

On May 23rd 2015, a neurologist named Mario Emilio Dourado who was the first to bring attention to the increasing number of Gullian-Barre Syndrome cases in the state of Rio Grande Do Norte, Natal City, Brazil. He described seven cases who presented with Zika virus disease and subsequently developed Gullian-Barre Syndrome.[60] As of January 2017, according to the WHO report, there have been approximately 18 countries with reported increases in incidence of Gullian-Barre Syndrome cases with at least one Gullian-Barre Syndrome case associated with confirmed Zika virus infection, and some countries with no increase in Gullian-Barre Syndrome have still reported at least one case associated with confirmed Zika virus infections.[54] Some of these countries include Brazil, Colombia, Puerto Rico, Dominican Republic, El Salvador, French Guiana, Bolivia, and Haiti.[60–64] This condition was also deemed a Public Health Emergency of International Concerns by the WHO director general in February 2014.[53] In lieu of the neurological complications being caused by Zika virus, the WHO released interim Guidance: Identification and Management of Guillian-Barre syndrome in the context of Zika virus disease, which was originally published on February of 2016 and later updated in August of 2016. These guidelines provided clinicians knowledge about how to approach, manage, and treat suspected cases of Gullian-Barre Syndrome associated with Zika virus disease.[65]

Cases of Zika virus disease started emerging in all South American countries and it seemed that the virus infection was advancing north toward the United States. It had spread to Venezuela, Colombia, Guyana, Suriname, Panama, Nicaragua, El Salvador, Honduras, Guatemala, Haiti, Cuba, Jamaica, and as far north as Mexico (Fig. 1.10).[66–70] This news spread panic in southern states like Florida and Texas, sparking fear that they would be the next target of this new viral pandemic. All these outbreaks had occurred in the geographical distribution of mosquito vectors, specifically the *Aedes aegypti* and *Aedes albopictus* mosquito. The global expansion of these arboviruses is most certainly linked to the spread of their vectors.[71]

ZIKA VIRUS DISEASE IN NORTH AMERICA

As the fear of Zika virus infection advanced North toward the United States of America, the media warned about potential cases of this virus infection in southern states.[72] Most notably, Florida and Texas were the first to theoretically be at risk. The geographical and ecological conditions in numerous southern states were ideal for the two main vectors of this virus.[71] It had been estimated by the CDC that, due to the vast geographical distribution of *Aedes aegpyti* and *Aedes albopictus* mosquitos in the United States reaching as far north as New York, the virus might not remain just in the southern states (Fig. 1.11).[73]

Although back in 2011 it had been reported that a scientist who returned from Senegal was infected with Zika virus and sexually transmitted the virus to his wife, this would be the first case of Zika virus reported in United States but it would not be the first case of the current outbreak.[30] The first reported case in the United States of Zika virus infection was identified in Harris County, Texas in January of 2016. It was identified in a traveler who was returning from South America where Zika virus disease was pres-

FIG. 1.10 South American countries where Zika virus disease has been reported up to 2016.

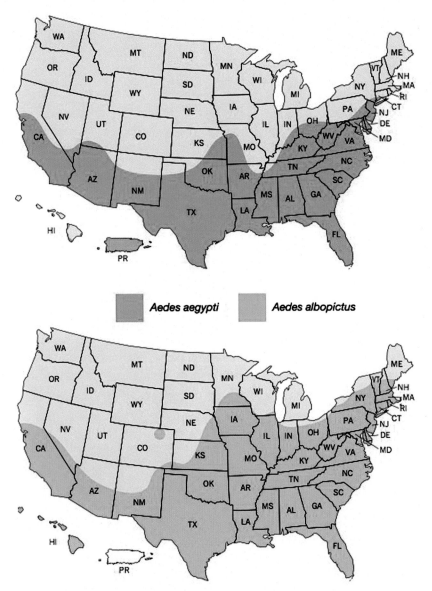

FIG. 1.11 Estimated range of *Aedes albopictus* and *Aedes aegypti* in United States. *Source: CDC (https://www.cdc.gov/zika/vector/range.html).*

ent, and this case would go on to be the first recorded case in the United States.[74] This prompted CDC to issue Interim Guidelines for pregnant women during this outbreak. The guidelines outlined recommendations for screening, testing, and management for any and all pregnant women who were exposed to the virus.[75]

In February of 2016, another report emerged of a potential Zika virus disease transmission by sexual contact in Dallas, Texas. A person who had traveled to a region of Zika virus disease outbreak had returned to the United States and had sexually transmitted Zika virus to their partner. Although there were reports of Zika-infected individuals in the United States, they were all travelers and this would be the first Zika virus infection on United States soil.[76] By February 16, 2016, a total of 116 residents of 33 different states and the District of Columbia had laboratory reported evidence of Zika virus disease. Out of these 116 cases, 115 were traveling associated infection cases.[77]

In early February in the state of Florida, Governor Rick Scott declared a public health emergency in four counties which included Miami-Dade in south Florida, Hillsborough in Tampa Bay region, Lee County in Southwest Florida, and Santa Rosa County in Florida Panhandle.[78] One of the very first travel warnings was issued by the CDC was on August 1, 2016 for people who were living or traveling to a 1 square mile of Wynwood neighborhood in Miami, Florida. In this warning the Florida Department of Health had identified an area with potential Zika virus transmission.[79] In this warning, they had various recommendations for pregnant women who were either living in or had traveled to this particular area and how they could avoid being affected. This included use of barrier contraception because Zika virus had the potential to spread via sexual intercourse.[80–83] These warnings were issued because the CDC was informed by Florida health officials in July, 2016 of four cases of Zika virus infections. These four cases would be the first local mosquito transmitted cases of Zika virus infection in the continental United States.[84] In December 14, 2016 CDC issued one of their most recent travel warning for those traveling to Brownsville, Cameron County, Texas.[85] This warning was due to Texas health officials reporting the first case of local mosquito-borne Zika virus disease in that area.[86]

The CDC has also very recently awarded $180 million to help assist Florida and Texas in the fight against Zika virus disease. This funding is for states, territories, local jurisdictions, and universities to support efforts to protect citizens from this virus. This was part of the $350 million in funding provided to CDC under the Zika Response and preparedness appropriation act of 2016. The efforts were for the following[87]:

1. Public Health Emergency Preparedness and Response to Zika virus disease related Activities
2. Zika virus disease Epidemiology and Laboratory Capacity Activities:
3. Zika virus related Birth Defects Surveillance Activities
4. Vector-Borne Disease Regional Centers of Excellence development
5. Vector Control Unit development in Puerto Rico

As of January 11, 2017, there have been a total of 216 locally acquired mosquito-borne reported cases of Zika virus disease, one laboratory acquired case, and 4649 travel associated cases. Out of these 4866 cases, there were 38 sexually

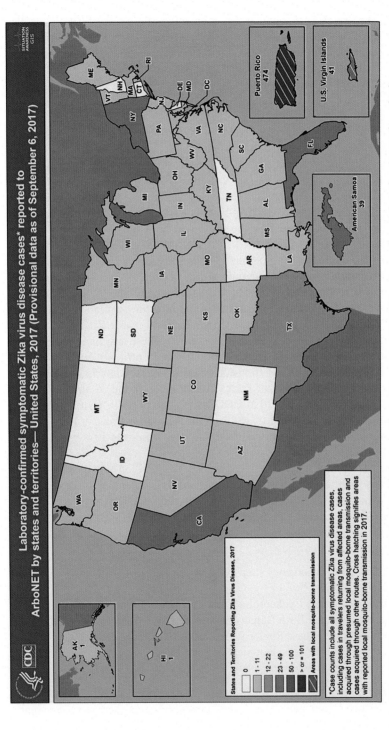

FIG. 1.12 Zika Virus disease cases in the United States According to State. *Source: CDC (https://www.cdc.gov/zika/intheus/maps-zika-us.html).*

transmitted cases and 12 cases of Guillain-Barre syndrome reported. All 50 states had reported laboratory confirmed Zika virus infections with Florida and New York having highest percentage (18% and 21%, respectively) of reported cases (Fig. 1.12). In the United States territories (American Samoa, US Virgins Islands, Puerto Rico), there have been 35,280 locally acquired cases out of which 97% occurred in Puerto Rico.[88]

REFERENCES

1. Prevention CfDCa. About Zika 2016 [cited 2016 01/05/2017]. Available from: https://www.cdc.gov/zika/about/index.html.
2. World Health Organization. The History of Zika Virus online. WHO, 2016. Available from: http://www.who.int/emergencies/zika-virus/timeline/en/.
3. Dick GW, Kitchen SF, Haddow AJ. Zika virus. I. Isolations and serological specificity. *Trans R Soc Trop Med Hyg* 1952;**46**(5):509–20.
4. Kaddumukasa MA, Mutebi JP, Lutwama JJ, Masembe C, Akol AM. Mosquitoes of Zika Forest, Uganda: species composition and relative abundance. *J Med Entomol* 2014;**51**(1):104–13.
5. Haddow AJ. The mosquitoes of Bwamba County, Uganda; mosquito breeding in plant axils. *Bull Entomol Res* 1948;**39**(Pt 2):185–212.
6. Dick GW. Zika virus. II. pathogenicity and physical properties. *Trans R Soc Trop Med Hyg* 1952;**46**(5):521–34.
7. Smithburn KC. Neutralizing antibodies against certain recently isolated viruses in the sera of human beings residing in East Africa. *J Immunol (Baltimore, Md: 1950)* 1952;**69**(2):223–34.
8. Macnamara FN. Zika virus: a report on three cases of human infection during an epidemic of jaundice in Nigeria. *Trans R Soc Trop Med Hyg* 1954;**48**(2):139–45.
9. Bearcroft WG. Zika virus infection experimentally induced in a human volunteer. *Trans R Soc Trop Med Hyg* 1956;**50**(5):442–8.
10. Boorman JP, Porterfield JS. A simple technique for infection of mosquitoes with viruses; transmission of Zika virus. *Trans R Soc Trop Med Hyg* 1956;**50**(3):238–42.
11. Kokernot RH, Casaca VM, Weinbren MP, McIntosh BM. Survey for antibodies against arthropod-borne viruses in the sera of indigenous residents of Angola. *Trans R Soc Trop Med Hyg* 1965;**59**(5):563–70.
12. Bres P. Recent data from serological surveys on the prevalence of arbovirus infections in Africa, with special reference to yellow fever. *Bull World Health Organ* 1970;**43**(2):223–67.
13. Robin Y, Bres P, Lartigue JJ, et al. Arboviruses in Ivory Coast. Serologic survey in the human population. *Bull Soc Pathol Exot Filiales* 1968;**61**(6):833–45.
14. Moore DL, Causey OR, Carey DE, et al. Arthropod-borne viral infections of man in Nigeria, 1964-1970. *Ann Trop Med Parasitol* 1975;**69**(1):49–64.
15. Simpson DI. Zika virus infection in man. *Trans R Soc Trop Med Hyg* 1964;**58**:335–8.
16. Fagbami AH. Zika virus infections in Nigeria: virological and seroepidemiological investigations in Oyo State. *J Hyg* 1979;**83**(2):213–9.
17. Marchette NJ, Garcia R, Rudnick A. Isolation of Zika virus from Aedes aegypti mosquitoes in Malaysia. *Am J Trop Med Hyg* 1969;**18**(3):411–5.
18. Olson JG, Ksiazek TG, Suhandiman, Triwibowo. Zika virus, a cause of fever in Central Java, Indonesia. *Trans R Soc Trop Med Hyg* 1981;**75**(3):389–93.
19. Hammon WM, Schrack Jr. WD, Sather GE. Serological survey for a arthropod-borne virus infections in the Philippines. *Am J Trop Med Hyg* 1958;**7**(3):323–8.

20. Darwish MA, Hoogstraal H, Roberts TJ, Ahmed IP, Omar F. A sero-epidemiological survey for certain arboviruses (Togaviridae) in Pakistan. *Trans R Soc Trop Med Hyg* 1983;**77**(4):442–5.
21. Smithburn KC, Kerr JA, Gatne PB. Neutralizing antibodies against certain viruses in the sera of residents of India. *J Immunol (Baltimore, Md: 1950)* 1954;**72**(4):248–57.
22. Korhonen EM, Huhtamo E, Smura T, et al. Zika virus infection in a traveller returning from the Maldives, June 2015. *Euro Surveill* 2016;**21**(2).
23. Buathong R, Hermann L, Thaisomboonsuk B, et al. Detection of Zika virus infection in Thailand, 2012-2014. *Am J Trop Med Hyg* 2015;**93**(2):380–3.
24. Roth A, Mercier A, Lepers C, et al. Concurrent outbreaks of dengue, chikungunya and Zika virus infections—an unprecedented epidemic wave of mosquito-borne viruses in the Pacific 2012-2014. *Euro Surveill* 2014;**19**(41).
25. Duffy MR, Chen TH, Hancock WT, et al. Zika virus outbreak on Yap Island, Federated States of Micronesia. *N Engl J Med* 2009;**360**(24):2536–43.
26. Oehler E, Watrin L, Larre P, et al. Zika virus infection complicated by Guillain-Barre syndrome—case report, French Polynesia, December 2013. *Euro Surveill* 2014;**19**(9).
27. Cao-Lormeau VM, Roche C, Teissier A, et al. Zika virus, French polynesia, South pacific, 2013. *Emerg Infect Dis* 2014;**20**(6):1085–6.
28. Mallet HP, Vial AL, Musso D. BISES: Bulletin d'information sanitaires, épidémiologiques et statistiques: Papeete: Bureau de veille sanitaire (BVS) Polynésie française, 2015. Available from: http://www.hygiene-publique.gov.pf/IMG/pdf/no13_-_mai_2015_-_zika.pdf.
29. Musso D, Roche C, Robin E, et al. Potential sexual transmission of Zika virus. *Emerg Infect Dis* 2015;**21**(2):359–61.
30. Foy BD, Kobylinski KC, Chilson Foy JL, et al. Probable non-vector-borne transmission of Zika virus, Colorado, USA. *Emerg Infect Dis* 2011;**17**(5):880–2.
31. Dupont-Rouzeyrol M, O'Connor O, Calvez E, et al. Co-infection with Zika and dengue viruses in 2 patients, New Caledonia, 2014. *Emerg Infect Dis* 2015;**21**(2):381–2.
32. Musso D, Nilles EJ, Cao-Lormeau VM. Rapid spread of emerging Zika virus in the Pacific area. *Clin Microbiol Infect* 2014;**20**(10):O595–6.
33. Dupont-Rouzeyrol M, Aubry M, O'Connor O, et al. Epidemiological and molecular features of dengue virus type-1 in New Caledonia, South Pacific, 2001-2013. *Virol J* 2014;**11**:61.
34. World Health Organization. Pacific Syndromic Surveillance report online. WHO, 2014. Available from: http://www.wpro.who.int/southpacific/programmes/communicable_diseases/disease_surveillance_response/PSS-30-Mar-2013/en/.
35. Besnard M, Lastere S, Teissier A, Cao-Lormeau V, Musso D. Evidence of perinatal transmission of Zika virus, French Polynesia, December 2013 and February 2014. *Euro Surveill* 2014;**19**(13).
36. Zanluca C, Melo VC, Mosimann AL, et al. First report of autochthonous transmission of Zika virus in Brazil. *Mem Inst Oswaldo Cruz* 2015;**110**(4):569–72.
37. Campos GS, Bandeira AC, Sardi SI. Zika virus outbreak, Bahia, Brazil. *Emerg Infect Dis* 2015;**21**(10):1885–6.
38. Heukelbach J, Alencar CH, Kelvin AA, de Oliveira WK. Pamplona de Goes Cavalcanti L. Zika virus outbreak in Brazil. *J Infect Dev Ctries* 2016;**10**(2):116–20.
39. Zika virus outbreaks in the Americas. Wkly Epidemiol Rec 2015;90(45):609–610.
40. Cardoso CW, Paploski IA, Kikuti M, et al. Outbreak of exanthematous illness associated with Zika, Chikungunya, and Dengue Viruses, Salvador, Brazil. *Emerg Infect Dis* 2015;**21**(12):2274–6.
41. Calvet GA, Filippis AM, Mendonca MC, et al. First detection of autochthonous Zika virus transmission in a HIV-infected patient in Rio de Janeiro, Brazil. *J Clin Virol* 2016;**74**:1–3.

42. Hennessey M, Fischer M, Staples JE. Zika virus spreads to new areas—region of the Americas, May 2015-January 2016. *MMWR Morb Mortal Wkly Rep* 2016;**65**(3):55–8.

43. Musso D. Zika virus transmission from French Polynesia to Brazil. *Emerg Infect Dis* 2015;**21**(10):1887.

44. Petersen E, Wilson ME, Touch S, et al. Rapid spread of Zika virus in The Americas—implications for public health preparedness for mass gatherings at the 2016 Brazil Olympic games. *Int J Infect Dis* 2016;**44**:11–5.

45. Triunfol M. A new mosquito-borne threat to pregnant women in Brazil. *Lancet Infect Dis* 2016;**16**(2):156–7.

46. Schuler-Faccini L, Ribeiro EM, Feitosa IM, et al. Possible association between Zika virus infection and microcephaly—Brazil, 2015. *MMWR Morb Mortal Wkly Rep* 2016;**65**(3):59–62.

47. Oliveira Melo AS, Malinger G, Ximenes R, et al. Zika virus intrauterine infection causes fetal brain abnormality and microcephaly: tip of the iceberg? *Ultrasound Obstet Gynecol* 2016;**47**(1):6–7.

48. Staples JE, Dziuban EJ, Fischer M, et al. Interim guidelines for the evaluation and testing of infants with possible congenital Zika virus infection—United States, 2016. *MMWR Morb Mortal Wkly Rep* 2016;**65**(3):63–7.

49. McCarthy M. CDC updates Zika virus guidance to protect pregnant women. *BMJ* 2016;**352**: . :i786.

50. Oduyebo T, Petersen EE, Rasmussen SA, et al. Update: interim guidelines for health care providers caring for pregnant women and women of reproductive age with possible Zika virus exposure—United States, 2016. *MMWR Morb Mortal Wkly Rep* 2016;**65**(5):122–7.

51. Vouga M, Musso D, Van Mieghem T, Baud D. CDC guidelines for pregnant women during the Zika virus outbreak. *Lancet* 2016;**387**(10021):843–4.

52. Prevention CfDCa. Pregnant Women: how to protect yourself CDC website2016. Available from: https://www.cdc.gov/zika/pregnancy/protect-yourself.html.

53. World Health Organization. WHO statement on the first meeting of the International Health Regulations (2005) (IHR 2005) Emergency Committee on Zika virus and observed increase in neurological disorders and neonatal malformations Internet. WHO, 2016. Available from: http://www.who.int/mediacentre/news/statements/2016/1st-emergency-committee-zika/en/.

54. World Health Organization. Zika Virus Microcephaly Guillain-Barre Syndrome online. WHO, 2017 [updated 5th January 2017]. Available from: http://apps.who.int/iris/bitstream/10665/252762/1/zikasitrep5Jan17-eng.pdf.

55. Jasti AK, Selmi C, Sarmiento-Monroy JC, et al. Guillain-Barre syndrome: causes, immunopathogenic mechanisms and treatment. *Expert Rev Clin Immunol* 2016;**12**(11):1175–89.

56. van den Berg B, Walgaard C, Drenthen J, et al. Guillain-Barre syndrome: pathogenesis, diagnosis, treatment and prognosis. *Nat Rev Neurol* 2014;**10**(8):469–82.

57. Solomon T, Dung NM, Vaughn DW, et al. Neurological manifestations of dengue infection. *Lancet* 2000;**355**(9209):1053–9.

58. Wielanek AC, Monredon JD, Amrani ME, Roger JC, Serveaux JP. Guillain-Barre syndrome complicating a Chikungunya virus infection. *Neurology* 2007;**69**(22):2105–7.

59. Wise J. Study links Zika virus to Guillain-Barre syndrome. *BMJ* 2016;**352**:i1242.

60. Araujo LM, Ferreira ML, Nascimento OJ. Guillain-Barre syndrome associated with the Zika virus outbreak in Brazil. *Arq Neuropsiquiatr* 2016;**74**(3):253–5.

61. Brasil P, Sequeira PC, Freitas AD, et al. Guillain-Barre syndrome associated with Zika virus infection. *Lancet* 2016;**387**(10026):1482.

62. Parra B, Lizarazo J, Jimenez-Arango JA, et al. Guillain-Barre syndrome associated with Zika virus infection in Colombia. *N Engl J Med* 2016;**375**(16):1513–23.

63. Dirlikov E, Ryff KR, Torres-Aponte J, et al. Update: Ongoing Zika Virus Transmission—Puerto Rico, November 1, 2015-April 14, 2016. *MMWR Morb Mortal Wkly Rep* 2016;**65**(17):451–5.

64. Paploski IA, Prates AP, Cardoso CW, et al. Time lags between exanthematous illness attributed to Zika virus, Guillain-Barre syndrome, and microcephaly, Salvador, Brazil. *Emerg Infect Dis* 2016;**22**(8):1438–44.

65. World Health Organization. Identification and management of Guillain-Barré syndrome in the context of Zika virus Interim guidance online. WHO, 2016 [updated 22 August 2016]. Available from: http://www.who.int/csr/resources/publications/zika/guillain-barre-syndrome/en/.

66. Valero N. Zika virus: another emerging arbovirus in Venezuela? *Investig Clin* 2015;**56**(3):241–2.

67. Sabogal-Roman JA, Murillo-Garcia DR, Yepes-Echeverri MC, et al. Healthcare students and workers' knowledge about transmission, epidemiology and symptoms of Zika fever in four cities of Colombia. *Travel Med Infect Dis* 2016;**14**(1):52–4.

68. Fabrizius RG, Anderson K, Hendel-Paterson B, et al. Guillain-Barre syndrome associated with Zika virus infection in a traveler returning from Guyana. *Am J Trop Med Hyg* 2016;**95**(5):1161–5.

69. Dos Santos T, Rodriguez A, Almiron M, et al. Zika virus and the guillain-barre syndrome—case series from seven countries. *N Engl J Med* 2016;**375**(16):1598–601.

70. Haque U, Ball JD, Zhang W, Khan MM, Trevino CJ. Clinical and spatial features of Zika virus in Mexico. *Acta Trop* 2016;**162**:5–10.

71. Kraemer MU, Sinka ME, Duda KA. The global distribution of the arbovirus vectors Aedes aegypti and Ae. albopictus. *elife* 2015;**4**:e08347.

72. Sun LH, Harwell D, Dennis B. Zika fear prompts travel warning for Miami, CDC's first in U.S. The Washington Post. 2016 August 1, 2016;Sect. Health & Science.

73. World Health Organization. Potential range in US online. WHO, 2016 [cited 2016]. Estimated range of Aedes albopictus and Aedes aegypti in the United States]. Available from: https://www.cdc.gov/zika/vector/range.html.

74. McCarthy M. First US case of Zika virus infection is identified in Texas. *BMJ* 2016;**352**:i212.

75. Petersen EE, Staples JE, Meaney-Delman D, et al. Interim guidelines for pregnant women during a Zika virus outbreak—United States, 2016. *MMWR Morb Mortal Wkly Rep* 2016;**65**(2):30–3.

76. McCarthy M. Zika virus was transmitted by sexual contact in Texas, health officials report. *BMJ* 2016;**352**:i720.

77. Armstrong P, Hennessey M, Adams M, et al. Travel-Associated Zika Virus Disease Cases Among U.S. Residents—United States, January 2015-February 2016. *MMWR Morb Mortal Wkly Rep* 2016;**65**(11):286–9.

78. Stein L. Florida governor declares health emergency in four counties over Zika online. Reuters; 2016. Available from: http://www.reuters.com/article/us-health-zika-florida-idUSKCN0VC2S9.

79. Prevention CfDCa. CDC guidance for travel and testing of pregnant women and women of reproductive age for zika virus infection related to the investigation for local Mosquito-borne Zika virus transmission in Miami-Dade and Broward Counties, Florida online. CDC; 2016 [updated August 1, 2016]. Available from: https://emergency.cdc.gov/han/han00393.asp.

80. Russell K, Hills SL, Oster AM, et al. Male-to-female sexual transmission of Zika virus-United States, January-April 2016. *Clin Infect Dis* 2017;**64**(2):211–3.

81. Oster AM, Brooks JT, Stryker JE, et al. Interim guidelines for prevention of sexual transmission of Zika virus. *MMWR Morb Mortal Wkly Rep* 2016;**65**(5):120–1.

82. Moreira J, Peixoto TM, Machado de Siqueira A, Lamas CC. Sexually acquired Zika virus: a systematic review. *Clin Microbiol Infect* 2017;**23**:296–305.

83. Tang WW, Young MP, Mamidi A, et al. A mouse model of zika virus sexual transmission and vaginal viral replication. *Cell Rep* 2016;**17**(12):3091–8.
84. Prevention CfDCa. Florida investigation links four recent Zika cases to local mosquito-borne virus transmission online. CDC, 2016 [updated July 29, 2016]. Available from: https://www.cdc.gov/media/releases/2016/p0729-florida-zika-cases.html.
85. Prevention CfDCa. Advice for people living in or traveling to Brownsville, Texas online. CDC, 2016 [updated December 14, 2016]. Available from: https://www.cdc.gov/zika/intheus/texas-update.html.
86. Prevention CfDCa. CDC supporting Texas investigation of possible local Zika transmission online. CDC, 2016 [updated November 28, 2016]. Available from: https://www.cdc.gov/media/releases/2016/p1128-zika-texas.html.
87. Prevention CfDCa. CDC awards nearly $184 million to continue the fight against Zika. CDC, 2016 [updated December 22, 2016]. Available from: https://www.cdc.gov/media/releases/2016/p1222-zika-funding.html.
88. Prevention CfDCa. Case Counts in the US online. CDC, 2017 [updated January 12, 2017]. US state and territories Zika Virus case counts.]. Available from: https://www.cdc.gov/zika/geo/united-states.html.

Chapter 2

Mosquito-Borne Diseases

Chapter Outline

MOSQUITO-BORNE DISEASES

Mosquitoes are considered one of the most dangerous species on the planet because they have the ability to spread many deadly diseases. Almost 700 million people contract a mosquito-borne illness every year resulting in greater than one million deaths.[1] The United State Centers for Disease Control and Prevention (CDC) reported that mosquitoes kill more than one million people a year just from the transmission of malaria. In recent years, the rate of infection spread has increased dramatically, and a growing number of scientists are concerned because global warming could lead to the explosive growth of the mosquito-borne diseases worldwide. These are some of the most common diseases spread around the world by mosquito bites: Zika virus, West Nile virus, and Chikungunya virus infections, Dengue fever, and malaria. There is no vaccine to prevent or medicine to treat most of these diseases.[2]

WEST NILE VIRUS INFECTION

West Nile virus is an arbovirus of the *Flavivirus* kind in the family *Flaviviridae*.[3] The virus can cause primarily an infection found in birds, but can also be transferred to humans, dogs, horses, and other animals. This virus first appeared in the U.S. in 1999 and spread by 2001 to Florida.

Zika Virus Disease. https://doi.org/10.1016/B978-0-12-812365-2.00003-2

This virus is fatal in some birds, especially crows and blue jays. Horses are also susceptible, but there is a vaccine available for disease prevention in horses. Humans, while also susceptible, often show no symptoms (less than 1% of those infected have severe symptoms). The mortality rate of those who manifest clinical symptoms ranges from 3% to 15%, and mortality is highest among the elderly persons. It is assumed that the main mosquito responsible for the transmission of human disease is one or more of the Culex species, of which there are at least eight species in Polk County of Florida, near the city of Lake Wales. Fifty-six mosquito species have been identified as carriers of West Nile virus across the United States.

Mosquitoes spread West Nile virus infection in 48 of the 50 U.S. states, Africa, Europe, the Middle East, and West and Central Asia. In 2012, the United States experienced one of its worst epidemics where 286 people died, with the maximum burden of disease in state of Texas.[4] As of 2014, there have been 36,437 cases of West Nile virus infection. Of these, 15,774 have resulted in meningitis/encephalitis and 1538 were fatal. There have been at least 1.5 million infections (82% are asymptomatic) and over 350,000 cases of West Nile virus infection, but the disease is under-reported due to its similarity to other viral infections. This virus usually circulates between mosquitoes and birds in Africa and Europe. However, in 1999 an outbreak of West Nile viral encephalitis was reported in New York City. Since then, the virus has spread to 48 states and the District of Columbia.[5-9]

West Nile virus infection is predominantly asymptomatic. About one in five infected individuals will have a fever and other flu-like symptoms. A few people (less than 1%) get more serious infections such as West Nile neuroinvasive disease (WNND), which causes meningitis, encephalitis, meningoencephalitis, and poliomyelitis-like syndrome. People of advanced age, the very young, or those with immunosuppression are most susceptible. Many patients with WNND have normal neuroimaging studies, although abnormalities may be present in various cerebral areas including the basal ganglia, thalamus, cerebellum, and brainstem.

West Nile virus encephalitis is the most common neuroinvasive manifestation of WNND. West Nile virus encephalitis presents with similar symptoms to others viral encephalitis including fever, headaches, and altered mental status. A prominent finding in WNE is the muscular weakness (30% to 50% of patients with encephalitis), often with lower motor neuron symptoms, flaccid paralysis, and hyporeflexia with no sensory abnormalities. West Nile meningitis (WNM) usually involves fever, headache, and stiff neck. Pleocytosis, an increase of white blood cells in cerebrospinal fluid, is also present. Changes in consciousness are not usually seen and are mild when present. West Nile poliomyelitis, an acute flaccid paralysis syndrome associated with infection, is less common than West Nile myelitis or encephalitis. This syndrome is generally characterized by the acute onset of asymmetric limb weakness or paralysis in the absence of sensory loss. The pain sometimes precedes the paralysis. The paralysis can occur in the absence of fever, headache, or other common symptoms associated with West Nile virus infection. Involvement of respiratory muscles, leading to acute respiratory failure, can sometimes occur. In West Nile reversible paralysis, like WNP, the weakness or paralysis is asymmetric.[10-15] Reported cases have an initial preservation of deep tendon reflexes, which is not expected for a pure anterior horn involvement. The disconnect of the upper motor neuron influences on the anterior horn cells possibly by myelitis or glutamate excitotoxicity have been suggested. The prognosis for recovery is excellent.[16]

Nonneurologic complications of West Nile virus infection include fulminant hepatitis, pancreatitis,[6] myocarditis, rhabdomyolysis,[13] orchitis,[15] optic neuritis,[10] cardiac dysrhythmias, and hemorrhagic fever with coagulopathy,[17] chorioretinitis.[18,19] The pregnant women and their babies may be at risk because prenatal infection has been linked to a birth defect called microcephaly.[11] There is no specific treatment for West Nile virus infection.[20,21]

CHIKUNGUNYA VIRUS INFECTION

Chikungunya is another arbovirus that belongs to the alphavirus genus of the family *Togaviridae* and is transmitted through the bite of an infected female mosquito belonging to *Aedes aegypti* and *Aedes albopictus*, two species that can also transmit other mosquito-borne viruses, including Dengue fever. The mosquitoes can bite during the day and at night.[22-24]

Chikungunya virus infection is a mosquito-borne viral disease first described during an outbreak in southern Tanzania in 1952. The name comes from an African language and refers to the stooped appearance people may have because of severe joint pain. Chikungunya virus infection has been identified in over 60 countries in Asia, Africa, Europe, and the Americas. The disease has established itself in the Caribbean (approximately 350,000 suspected cases in the Americas since December 2013). Two cases of locally transmitted Chikungunya virus infection have been identified in Florida in July of 2014. As of July 22, 2014, 497 travel-related cases have been found in 35 states, Puerto Rico, and the U.S. Virgin Islands.

Chikungunya virus infection has recently appeared in such places as India, Sri Lanka, Mauritius, and other countries in Europe involved in frequent tourism to these destinations. An increase in geographical areas involved in Europe is expected due to the spread of the Asian Tiger mosquitoes (*Aedes albopictus*) that can act as a vector for this infection. The traditional areas of involvement for this virus also include Africa and Southeast Asia.

- Symptoms usually appear 3–7 days after the bite of an infected mosquito. The most common symptoms of Chikungunya virus infection are fever (abrupt onset of fever above 38°C lasting 2 to 3 days) and joint pain (moderate-to-intense joint and tendon pain and swelling). Other symptoms may include a headache, muscle pain, joint swelling, or rash. Chikungunya virus disease does not often result in death, but the symptoms can be severe and disabling. Most patients recover fully, but in some cases, joint pain may persist for several months or even years. Rash (between the 2nd and 5th day in 50% of the cases) and conjunctivitis/eye redness (30% of the cases) have been reported.

- Very rare cases include neurological manifestations in the form of encephalitis and myelitis. Serious complications are not common except in older people where the disease can cause of death. Often symptoms in infected individuals are weak and the infection may go unrecognized or be misdiagnosed in areas where Dengue fever occurs.[25-31]

No specific treatment for chikungunya virus infection is available at this stage. The infected person should take analgesics to treat fever (except acetylsalicylic acid) and antiinflammatory for the joint pain, ensure appropriate hydration and nutrition, and take precautions against mosquito biting to prevent spread.[32]

DENGUE VIRUS INFECTION

Dengue virus is an arbovirus that spreads through the bite of an infected female mosquito of the *Aedes* genus, chiefly the *Ae. aegypti*. Nearly unique among arboviruses, the Dengue viruses utilize humans as their only natural vertebrate host. The mosquitoes *Aedes aegypti* and *Ae .albopictus* are the principal vectors in most of the world; both are common in Florida.[33,34] A mosquito is able to transmit dengue virus for a week after biting an infected person. More than one-third of the world's population lives in areas at risk for infection, and Dengue virus infection is a leading cause of morbidity and mortality in the tropics and subtropics. As many as 400 million people are infected annually. Dengue virus infection has become a worldwide problem since the 1950s. Although Dengue virus infection rarely occurs in the continental United States, the disease is endemic in Puerto Rico and in many popular tourist destinations in Latin America, Southeast Asia, and the Pacific islands. In recent years, Dengue virus infection occurred in the southern states, including Texas and Florida. Dengue virus infection was first identified in Florida in 1850 and until 1934 the epidemic in Miami was estimated to include more than 15,000 cases.[33,35,36]

Dengue virus infection is a disease with symptoms ranging from simple flu-like illness to severe hemorrhagic symptoms, shock, encephalitis, or death, and is caused by any of four distinct Dengue virus species (DEN-1, -2, -3, or -4). There are four serotypes of the virus that causes Dengue fever and infection by one subtype provides lifelong immunity against that subtype but not against the others. As the Dengue virus multiplies and damages the cells, an infected person begins to show symptoms similar to other infections: high fever, headaches, back and joint pain, rashes, and eye pain. Symptoms of infection by Dengue fever characteristically include: fever above 38°C lasting 4–7 days, intense headache, retro-orbital pain, severe arthralgias and myalgias, and extreme fatigue lasting for days or weeks. A rash may develop after the 4th day in 30%–50%, and nausea, vomiting, and sometimes diarrhea may be seen in 30%–50% of the cases.

A severe and life-threatening Dengue hemorrhagic fever (DHF) may occasionally develop.[37-42,33,43,44,34,45-56] The fever lasts from 2 to 7 days, with general signs and symptoms consistent with Dengue fever. When the fever resolves, other signs such as abdominal pain, bleeding, vomiting of blood, blood in the feces, and epistaxis will develop. At the beginning of 24–48 h of Dengue fever, the smallest blood vessels (capillaries) become excessively permeable ("leaky"), resulting in fluid shift from the blood vessels into the peritoneum (causing ascites) and pleural cavity (leading to pleural effusions). This can lead to circulatory failure, and shock. In addition, the patients with Dengue hemorrhagic fever have low platelet count and hemorrhagic manifestations, tendency to bruise easily, or other types of skin hemorrhages, bleeding nose or gums, and possibly internal bleeding.[37–42,33,43,44,34,45–56] The fatality rate for hemorrhagic fever is about 5% and is more likely in children.[36]

In very rare cases, neurological complications, such as encephalitis, Guillain-Barré syndrome, and myelitis may be observed. Like most viruses, there is no specific treatment.[51,52,55]

EASTERN EQUINE ENCEPHALITIS VIRUS INFECTION

Eastern equine encephalitis virus is transmitted to humans by the bite of an infected mosquito. Eastern equine encephalitis is a rare illness in humans, and only a few cases are reported in the United States each year. This is a disease found in horses, but there is a vaccine available to prevent equine infection. The virus is maintained in birds, which are vectors but not affected by the disease. Humans do sometimes contract the disease, but humans are not a preferred target of the mosquitoes carrying this virus and are considered a "dead end" host. This means that, while a human can contract Eastern Equine encephalitis, mosquitoes have not been shown to become infected by a human host and spread it to other organisms. Several states in the Northeast United states have seen increased virus infections since 2004.[57] Most cases occur in the Atlantic and Gulf Coast states. Most persons infected with Eastern equine encephalitis virus have no apparent illness. Severe cases of Eastern equine encephalitis, involving encephalitis, begin with the sudden onset of a headache, high fever, chills, and vomiting. The illness may then progress into disorientation, seizures, or coma. Eastern Equine encephalitis is one of the most severe mosquito-transmitted diseases in the United States with approximately 33% mortality and significant brain damage in most survivors. There is no specific treatment for Eastern Equine encephalitis; care is based on symptoms.[58,59]

JAPANESE ENCEPHALITIS VIRUS

Japanese encephalitis virus infection is the leading cause of vaccine-preventable encephalitis in Asia and the Western Pacific.[60–62] The virus is transmitted to humans by the bite of an infected mosquito, which serves as a dead end host due to its short duration and low viremia in man. Pigs are an important maintenance host for this virus, which is mainly transmitted by night-biting mosquitoes in the

Culextritaeniorhynchus group. In the pigs, however, virus does not produce encephalitis, although abortion occurs in pregnant sows. Most important mosquito vector in Asia is *Culextritaeniorhynchus,* which breeds in the stagnant water like paddy fields or drainage ditches. Other species are *Culexvishnui* (India), *C. gelides, C. fusco cephalea* (India, Malaysia, Thailand), and *C. pipiens.* Countries with the proven epidemics of Japanese encephalitis virus are India, Pakistan, Nepal, Sri Lanka, Burma, Laos, Vietnam, Malaysia, Singapore, Philippines, Indonesia, China, maritime Siberia, Korea, and Japan. Since the 1990s the virus infection has spread in Pakistan, Nepal, and Australia. The first clinical case of Japanese encephalitis virus in India was observed in 1955 at Vellore (former North Arcot district, Tamil Nadu). A total of about 65 cases were reported between 1955 and 1966 in South India. In 1973, the first major viral infection outbreak occurred in Burdwan and Bankura, the two districts of West Bengal with about 700 cases reported resulting in 300 deaths. Subsequently, another outbreak in the same state occurred in 1976 with 307 cases reported resulting in 126 deaths. Around 30,000 to 50,000 clinical infections are reported each year across Asia. The majority of infections, however, are asymptomatic.[63–67] There is an incubation period of 4–14 days in humans during Japanese encephalitis virus infection and patients present within a few days of infection with fever including coryza, diarrhea, and rigors. Convulsions occur in 10% of the infected patients, more frequently in children (85% of cases) than in adult patients (75% of cases). About 50%–60% of the survivors suffer from serious long-term neurologic complications manifesting as convulsions, tremors, paralysis, ataxia, memory loss, impaired cognition, behavioral disturbance, and other such symptoms. Thirty percent of those who demonstrated Japanese encephalitis virus symptoms die and another 30% develop serious and permanent neurological damage. Steps to prevent Japanese encephalitis virus include using personal protective measures to prevent mosquito bites.

Vector control alone cannot be relied upon to prevent Japanese encephalitis virus since it is practically almost impossible to control mosquito density in the rural areas which are the worst affected areas due to poor socio-economic conditions. As previously mentioned, an effective vaccine is available for preventing this infection.[68] Three types of Japanese encephalitis virus vaccine are currently in use: mouse-brain derived inactivated, cell-culture-derived inactivated, and cell-culture-derived live attenuated Japanese encephalitis virus vaccine. Formalin-inactivated vaccines are safe and effective against Japanese encephalitis virus infection and in use for at least 30 years.

MURRAY VALLEY ENCEPHALITIS VIRUS

Murray Valley encephalitis is an uncommon disease caused by the Murray Valley encephalitis virus. Murray Valley encephalitis virus is a flavivirus[69] endemic to northern Australia and Papua New Guinea.[70] The most important species of mosquito to carry the virus is the common banded mosquito, *Culexannulirostris.*

Most people with this infection remain completely asymptomatic, while others may only develop a mild illness with fever. A small proportion of those infected develop encephalitis.

Symptoms of Murray Valley encephalitis usually appear 5–28 days (average 14 days) after the infected mosquito bite. The early symptoms include headache, fever, nausea, vomiting, and muscle aches. Symptoms may also include drowsiness, confusion, or seizures (especially in infants), and in severe cases delirium, coma, and death. Some who recover are left with ongoing problems such as deafness or epilepsy. There is no specific treatment for Murray Valley encephalitis.[71–73]

LA CROSSE ENCEPHALITIS VIRUS

La Crosse encephalitis was discovered in 1965, after the virus was isolated from preserved brain tissue and spinal cord of a child who died from the unknown infection in La Crosse, Wisconsin in 1960. The viral infection occurs in the Appalachian and Midwestern regions of the United States. Recently there has been an increase of viral infection cases in the Southeast of the United States. An explanation to this may be that the mosquito *Aedes albopictus* is also an efficient vector of La Crosse virus.[74] Between 2004 and 2013, there were 787 total cases of La Crosse encephalitis and 11 deaths in the United States.[75]

Many people infected with La Crosse Encephalitis virus are asymptomatic. The incubation period for La Crosse Encephalitis virus disease ranges from 5 to 15 days. La Crosse Encephalitis virus disease is usually characterized by fever (usually lasting 2–3 days), headache, nausea, vomiting, fatigue, and lethargy. The severe form of La Crosse Encephalitis virus disease often results in encephalitis and can include seizures, coma, and paralysis. Severe disease occurs most often in children under the age of 16 years. In rare cases, long-term disability or death can result from La Crosse encephalitis. No vaccine against La Crosse Encephalitis virus infection or specific antiviral treatment for clinical La Crosse Encephalitis virus infection is available. Patients with suspected La Crosse Encephalitis virus encephalitis should be hospitalized, and appropriate serologic and other diagnostic tests ordered, and supportive treatment (including seizure control) provided.[76–78]

MALARIA INFECTION

Malaria is a disease transmitted by *Anopheles* mosquitoes in the genus distributed throughout the world as a mosquito-borne disease caused by a parasite. Malaria is present in more than 100 countries and imposes a significant burden in regards to financial cost and deaths involving at least 80 million people every year. Malaria kills an estimated 1.1 million people per year, and probably more deaths occur but are not reported due to incomplete case ascertainment in many of the countries with the greatest burdens. Four species of parasites affect humans, but two of them, *Plasmodium falciparum* and *Plasmodium vivax*, account for more than 95% of cases. Infection from *Plasmodium falciparum,* the more virulent one, is highly prevalent throughout the deep tropics from Africa to Asia and South America. In 2015, an estimated 214 million cases of malaria occurred worldwide and 438,000 people died, mostly children in the African region. Malaria was endemic in the United States in the 19th and 20th centuries. The disease was considered "eradicated" from the Unites States by mid-1950s. From 1957 through 1996, 78 human cases of malaria infection have been reported from 22 states. Three of the cases were in Florida since malaria was eradicated in the State in 1948, one was a woman camping in the panhandle's Gulf County in 1990, and two in men living in Palm Beach County in 1996. These patients had never lived or visited areas with malaria. About 1500 cases of malaria are diagnosed in the United States each year. The vast majority of cases in the United States are in travelers and immigrants returning from countries where malaria transmission occurs, many from sub-Saharan Africa and South Asia.[79–84]

The modifying structural and virulent characteristics of malaria parasites prevents complete immunity from developing, but older children and adults who have experienced multiple infections, experience some level immunity from the most severe manifestations of the illness.

The causative agent of tropical malaria-infected red blood cells, especially mature trophozoites, adhere to the vascular endothelium of venular blood vessel walls but do not freely circulate in the blood. When this sequestration of infected erythrocytes occurs in the vessels of the brain, it is believed to be a factor in causing the severe disease syndrome known as cerebral malaria, which is associated with high mortality.[85–88] Stories of patients falling ill on a Friday, putting off treatment till Monday, and dying over the weekend are not uncommon.

Although no vaccine is currently available, prophylactic drugs and measures that reduce exposure to night-biting *Anopheles* mosquitoes, such as bed nets and repellents can be very effective.

UNCOMPLICATED MALARIA

The classical (but rarely observed) malaria attack lasts 6–10 hours. It consists of a cold stage (sensation of cold, shivering), a hot stage (fever, headaches, vomiting; seizures in young children), and finally a sweating stage (sweats, return to normal temperature, tiredness). Classically (but infrequently observed) the attacks occur every second day with the "tertian" parasites (*Plasmodium falciparum, Plasmodium vivax,* and *Plasmodium ovale*) and every third day with the "quartan" parasite (*Plasmodium malaria*). More commonly, the patient presents with a combination of the following symptoms: fever, chills, sweats, headaches, nausea, vomiting, body aches, and general malaise.

The diagnosis of malaria depends on the demonstration of parasites in the blood, usually by microscopy. Additional laboratory findings may include mild anemia, the slight decrease in blood platelets (thrombocytopenia), elevation of bilirubin, and elevation of aminotransferases.[89,1,90–96]

SEVERE MALARIA

Severe malaria occurs when infections are complicated by serious organ failures or abnormalities in the patient's blood or metabolism. The manifestations of severe malaria include:

- Cerebral malaria, with abnormal behavior, impairment of consciousness, seizures, coma, or other neurologic abnormalities
- Severe anemia and hemoglobinuria due to hemolysis and abnormalities in blood coagulation
- Acute respiratory distress syndrome, an inflammatory reaction in the lungs that inhibits oxygen exchange, which may occur even after the parasite counts have decreased in response to treatment
- Low blood pressure caused by cardiovascular collapse
- Acute kidney failure
- Hyperparasitemia, where more than 5% of the red blood cells are infected by malaria parasites
- Metabolic acidosis, often in association with hypoglycemia
- Hypoglycemia, which may also occur in pregnant women with uncomplicated malaria, or after treatment with quinine
- Neurologic defects may occasionally persist following cerebral malaria, especially in children. Such defects include ataxia, palsies, speech difficulties, deafness, and blindness.[97–102]

Quinine and other antimalarial drugs cure patients by targeting the parasites in the blood. No effective vaccine exists at this point. The recommended treatment for malaria is the combination of antimalarial medications that includes an artemisinin. The second medication may be either mefloquine, lumefantrine, or sulfadoxine pyrimethamine. Quinine along with doxycycline may be used if artemisinin is not available. There are a number of drugs that can help prevent or interrupt malaria infection in travelers to areas where the infection is common. Many of these drugs are also used in treatment. Chloroquine may be used where chloroquine-resistant parasites are not common. In places where *Plasmodium* is resistant to one or more medications, three medications—mefloquine (*Lariam*), doxycycline (available generically), or the combination of atovaquone and proguanil hydrochloride (*Malarone*)—are frequently used when prophylaxis is needed. Doxycycline and the atovaquone plus proguanil combination are the best tolerated; mefloquine can rarely be associated with death, suicide, and neurological and psychiatric symptoms.[103–107] Drug resistance is now common against all classes of antimalarial drugs apart from artemisinins.[108]

ST. LOUIS ENCEPHALITIS VIRUS INFECTION

St. Louis encephalitis viral disease is maintained in birds and transmitted to both humans and horses by the mosquito *Culexnigrapalpus*. St. Louis encephalitis (SLE) is transmitted from birds to humans and other mammals by infected mosquitoes (mainly some *Culex* species). While the virus is consistently present in the wild every year, it causes a public health outbreak only periodically. This irregular infection outbreak pattern is the result of the number of susceptible individuals in the bird population coupled with locally residing large populations of the transmitting mosquito. These mosquito population increases are triggered by a combination of climactic conditions including drought and rain cycles and individual rain events.

Currently, the best method of determining the presence of active viral transmission is the use of sentinel chicken flocks. Coupled with mosquito population surveillance, the data provides a basis for identifying emergence of infection.

St. Louis encephalitis virus is found throughout the United States, but most often along the Gulf of Mexico, especially Florida. Major St. Louis encephalitis virus epidemics occurred in Florida in 1959, 1961, 1962, 1977, and 1990. The elderly and very young are more susceptible than those between 20 and 50 years. Between 1964 and 1998 (a 35-year period) a total of 4478 confirmed cases of St. Louis encephalitis virus were recorded in the United States. Symptoms are similar to those seen in Eastern equine encephalitis and like Eastern equine encephalitis and there is no vaccine. Mississippi's first case of St. Louis encephalitis virus since 1994 was confirmed in June 2003. Previously the last outbreak of St. Louis encephalitis virus in Mississippi was in 1975 with over 300 reported cases. It was the first confirmed mosquito-borne virus in the United States in 2003. It emerged in October 2003 in California Riverside County in sentinel chickens. The last St. Louis encephalitis virus human case in California occurred in 1997.[109–112]

Humans exposed to the virus are often asymptomatic. Less than 1% of St. Louis encephalitis virus infections are clinically apparent and the vast majority of infections remain undiagnosed. The incubation period for SLE ranges from 5 to 15 days. The onset of illness is usually abrupt, with fever, headache, dizziness, nausea, and malaise. Signs and symptoms intensify over a period of several days to a week. Some patients spontaneously recover after this period, and others develop signs of central nervous system infections, including stiff neck, confusion, disorientation, dizziness, tremors, and unsteadiness. Coma can develop in severe cases. The disease is generally milder in children than in older adults. About 40% of children and young adults with St. Louis encephalitis virus disease develop only fever and headache or aseptic meningitis. Almost 90% of elderly persons with St. Louis encephalitis virus disease develop encephalitis. The overall case-fatality ratio is 5% to 15%. The risk of fatal disease also increases with age.[110,113]

There is no vaccine or any other treatments specifically for Saint Louis encephalitis virus, although one study showed that early use of interferon-alpha2b may decrease the severity of complications.[114]

YELLOW FEVER VIRUS INFECTION

Yellow fever virus, which has a 400-year history, is found in tropical and subtropical areas in South America and Africa. Every year about 200,000 cases occur with 30,000 deaths in 33 countries. The disease has not been reported in Asia. Over the past decade, it has become more prevalent. Yellow fever is a very rare cause of illness in U.S. In 2002, one fatal yellow fever infection death occurred in the United States in an unvaccinated traveler returning from a fishing trip to the Amazon. Like Dengue fever, it is transmitted by *Aedes* mosquitoes, especially *Aedesaegypti*, the yellow fever mosquito.[115,116]

Most people infected with the yellow fever virus either have no manifestations or only mild illness. In persons who develop symptoms, the incubation period (time from infection until illness) is typically 3–6 days. The initial symptoms include sudden onset of fever, chills, severe headache, back pain, general body aches, nausea, vomiting, fatigue, and weakness. Most people improve after the initial presentation. After a brief remission of hours to a day, roughly 15% of cases progress to develop a more severe form of the disease. The severe form is characterized by high fever, jaundice that is why the disease is called yellow fever, bleeding, and eventually shock and failure of multiple organs.

Yellow fever disease is diagnosed based on symptoms, physical findings, laboratory testing, and travel history, including the possibility of exposure to infected mosquitoes.[117–120] There is no specific treatment for yellow fever. Vaccination is highly recommended as a preventive measure for travelers to and people living in endemic countries.[121]

RIFT VALLEY FEVER VIRUS INFECTION

First reported among livestock in Rift Valley of Kenya in the early 1900s,[122] and the virus was first isolated in 1931. Infected mosquitoes can give the disease to people and animals. It is common in parts of Africa. People have also contracted the virus infection in Saudi Arabia and Yemen.[123,124] Infection outbreaks usually occur during periods of increased rain, which increase the number of mosquitoes.[125] The virus is transmitted through mosquito vectors, as well as through contact with the tissue of infected animals. Two species— *Culextritaeniorhynchus* and *Aedes vexans*—are known to transmit the virus. The mild symptoms may include fever, muscle pains, and headaches which often last for up to a week.

In humans, the virus can cause several syndromes. Usually, infected persons may have either no symptoms or only a mild illness with fever, headache, muscle pains, and liver[126] abnormalities. Patients who become ill usually experience fever, generalized weakness, back pain, dizziness, and weight loss at the onset of the illness. Typically, people recover within 2–7 days after onset. In a small percentage of cases (<2%), the illness can progress to hemorrhagic fever syndrome, meningoencephalitis[127] (inflammation of the brain and tissues lining the brain), or affect the eye. The severe symptoms may include: loss of the ability to see (beginning three weeks after the infection), encephalitis that causes severe headaches and confusion, and hemorrhagic manifestations together with liver failure, which may occur within the first few days. Overall mortality rate is about 1% but may be as high as 50% in those who develop hemorrhagic manifestations. Other signs in livestock include vomiting and diarrhea, respiratory disease, fever, lethargy, anorexia, and sudden death in young animals.[128,126,129]

There is a human vaccine[126]; but, as of 2010, it is not widely available. There is no specific treatment and medical treatment is based on supportive care.[126,130]

KUNJIN VIRUS INFECTION

The main mosquito associated with the spread of Kunjin virus infection is *Culexannulirostris,* which is geographically widespread and is associated with fresh water habitats. The kunjin virus is a zoonotic virus from the family *Flaviviridae*[131] and the genus *Flavivirus*. The virus was isolated in mosquitoes in South East Asia but in humans, only in Australia.[132,133] Symptoms of Kunjin virus disease vary. The vast majority of infected people do not develop any symptoms. A small number of people may experience a mild illness with symptoms including fever, headache, muscle pain, swollen lymph nodes, fatigue, and rash. Some of those people may experience encephalitis. Symptoms of encephalitis may include confusion, drowsiness, and seizures. There is no specific treatment for the disease.[131,134,135]

ROSS RIVER VIRUS INFECTION

Ross River virus infection is endemic to Australia, Papua New Guinea, and other islands in the South Pacific.[136] Ross River virus is transmitted from animals to humans by a number of different types of mosquitoes with *Culexannulirostris, Aedes vigilax* (salt marsh mosquito) and *Aedes notoscriptus* being most common. Ross River virus infection causes epidemic polyarthritis. The symptoms may include fever with joint pain and swelling which may then be followed in 1 to 10 days by a raised erythematous rash affecting mainly the trunk and limbs. The rash usually lasts for 1 to 10 days and may or may not be accompanied by a fever. The joint pain can be severe and usually lasts 2 to 6 weeks.[137–139]

BARMAH FOREST VIRUS INFECTION

Barmah Forest virus is an *Alphavirus*. This disease was named after the area in Northern Victoria where it was first isolated in 1974. As of 2015, it has been found only in Australia.[140–142] Barmah Forest virus infection causes inflammation and joint pain and has similar symptoms to Ross River virus infection (epidemic polyarthritis), but usually, lasts for a shorter duration. The symptoms may include fever, headache, tiredness, painful joints, joint swelling, muscle tenderness, and skin rashes. Some people, especially children, may become infected without showing any symptoms. The initial fever and discomfort only last a few days but some people may experience joint pain, tiredness, and muscle tenderness for up to 6 months. There is no specific drug treatment for Barmah Forest virus infection.[141,143–145]

JAMESTOWN CANYON VIRUS INFECTION

Jamestown Canyon virus infection, which can be transmitted to several different species of mosquitoes throughout Minnesota, is a rarely reported cause of illness in humans.[146] The virus is closely related to La Crosse virus, although the disease is reported less frequently and any age group may be affected.

About 2 days to 2 weeks after the bite from an infected mosquito, disease symptoms are nonspecific summertime illness with a sore throat, runny nose, and cough, followed by fever, headache, nausea, and vomiting may develop. The neuroinvasive[147] disease occurs in two-thirds of reported cases and is characterized by a severe headache and neck stiffness as in meningitis or increasing lethargy and altered mental status up to coma as in meningoencephalitis. No specific therapy exists for arboviral infections; treatment is limited to supportive care and managing complications, such as relieving increased intracranial pressure.[147]

REFERENCES

1. Caraballo H. Emergency Department Management of Mosquito-Borne Illness: Malaria, Dengue, and West Nile Virus. *Emerg Med Pract* 2014;**16**(5):1–23.
2. Tolle MA. Mosquito-borne diseases. *Curr Probl Pediatr Adolesc Health Care* 2009;**39**(4):97–140.
3. Klenk K, Snow J, Morgan K, Bowen R, Stephens M, Foster F, et al. Alligators as West Nile virus amplifiers. *Emerg Infect Dis* 2004;**10**(12):2150–5.
4. Murray K, Ruktanonchai D, Hesalroad D, Fonken E, Nolan M. West Nile virus, Texas, USA, 2012. *Emerg Infect Dis* 2013;**19**(11):1836–8.
5. Asnis D, Conetta R, Teixeira A, Waldman G, Sampson B. The West Nile virus outbreak of 1999 in New York: The Flushing hospital experience. *Clin Infect Dis* 2000;**30**(3):413–8.
6. Calisher C. West Nile virus in the New World: appearance, persistence, and adaptation to a new econiche—an opportunity taken. *Viral Immunol* 2000;**13**(4):411–4.
7. Chen C, Jenkins E, Epp T, Waldner C, Curry P, Soos C. Climate change and West Nile virus in a highly endemic region of North America. *Int J Environ Res Public Health* 2013;**10**(7):3052–71.
8. Hayes E, Komar N, Nasci R, Montgomery S, O'Leary D, Campbell G. Epidemiology and transmission dynamics of West Nile virus disease. *Emerg Infect Dis* 2005;**11**(8):1167–73.
9. Nash D, Mostashari F, Fine A, et al. The outbreak of West Nile virus infection in the New York City area in 1999. *N Engl J Med* 2001;**344**(24):1807–14.
10. Anninger W, Lomeo M, Dingle J, Epstein A, Lubow M. West Nile virus-associated optic neuritis and chorioretinitis. *Am J Ophthalmol* 2003;**136**(6):1183–5.
11. Carson P, Konewko P, Wold K, et al. Long-term clinical and neuropsychological outcomes of West Nile virus infection. *Clin Infect Dis* 2006;**43**(6):723–30.
12. Davis L, DeBiasi R, Goade D, et al. West Nile virus neuroinvasive disease. *Ann Neurol* 2006;**60**(3):286–300.
13. Montgomery S, Chow C, Smith S, Marfin A, O'Leary D, Campbell G. Rhabdomyolysis in patients with west nile encephalitis and meningitis. *Vector Borne Zoonotic Dis* 2005;**5**(3):252–7.

14. Papa A, Karabaxoglou D, Kansouzidou A. Acute West Nile virus neuroinvasive infections: cross-reactivity with dengue virus and tick-borne encephalitis virus. *J Med Virol* 2011;**83**(10):1861–5.

15. Smith R, Konoplev S, DeCourten-Myers G, Brown T. West Nile virus encephalitis with myositis and orchitis. *Hum Pathol* 2004;**35**(2):254–8.

16. Mojumder D, Agosto M, Wilms H, et al. Is initial preservation of deep tendon reflexes in West Nile virus paralysis a good prognostic sign? *Neurol Asia* 2014;**19**(1):93–7.

17. Paddock C, Nicholson W, Bhatnagar J, et al. Fatal hemorrhagic fever caused by West Nile virus in the United States. *Clin Infect Dis* 2006;**42**(11):1527–35.

18. Shaikh S, Trese M. West Nile virus chorioretinitis. *Br J Ophthalmol* 2004;**88**(12):1599–600.

19. Fonseca K, Prince G, Bratvold J, et al. West Nile virus infection and conjunctive exposure. *Emerg Infect Dis* 2005;**11**(10):1648–9.

20. Klee A, Maidin B, Edwin B, et al. Long-term prognosis for clinical West Nile virus infection. *Emerg Infect Dis* 2004;**10**(8):1405–11.

21. Watson J, Pertel P, Jones R, et al. Clinical characteristics and functional outcomes of West Nile fever. *Ann Intern Med* 2004;**141**(5):360–5.

22. Chikungunya virus symptoms, diagnosis, & treatment. CDC, 2016. Retrieved 26 September 2016.

23. Thiberville S-D, Moyen N, Dupuis-Maguiraga L, Nougairede A, Gould E, Roques P, et al. Chikungunya fever: Epidemiology, clinical syndrome, pathogenesis and therapy. *Antiviral Res* 2013;**99**(3):345–70.

24. Staples J, Breiman R, Powers A. Chikungunya fever: an epidemiological review of a re-emerging infectious disease. *Clin Infect Dis* 2009;**49**:942–8.

25. Caglioti C, Lalle E, Castilletti C, Carletti F, Capobianchi MR, Bordi L. Chikungunya virus infection: an overview. *New Microbiol* 2013;**36**(3):211–27.

26. Fourie ED, Morrison JG. Rheumatoid arthritic syndrome after chikungunya fever. *South African Medical [Suid-Afrikaansetydskrifvirgeneeskunde]* 1979;**56**(4):130–2.

27. MacFadden D, Bogoch I. Chikungunya. *Can Med Assoc J* 2014;**186**(10):775.

28. Mahendradas P, Ranganna SK, Shetty R, Balu R, Narayana KM, Babu RB, et al. Ocular manifestations associated with chikungunya. *Ophthalmology* 2008;**115**(2):287–91.

29. Powers, A. Chikungunya. CDC. Retrieved 12 May 2014.

30. Schilte C, Staikowsky F, Staikovsky F, Couderc T, Madec Y, Carpentier F, et al. Chikungunya virus-associated long-term arthralgia: a 36-month prospective longitudinal study. *PLoS Negl Trop Dis* 2013;**7**(3). e2137.

31. Weaver SC, Lecuit M. Chikungunya Virus and the Global Spread of a Mosquito-Borne Disease. *N Engl J Med* 2015;**372**(13):1231–9.

32. Weaver SC, Osorio JE, Livengood JA, Chen R, Stinchcomb DT. Chikungunya virus and prospects for a vaccine. *Expert Rev Vaccines* 2012;**11**(9):1087–101.

33. Gubler D. Dengue viruses. In: Mahy BWJ, Van Regenmortel MHV, editors. *Desk encyclopedia of human and medical virology*. Boston: Academic Press; 2010. p. 372–82.

34. Henchal E, Putnak J. The dengue viruses. *Clin Microbiol Rev* 1990;**3**(4):376–96.

35. Amarasinghe A, Kuritsk JN, Letson GW, Margolis HS. Dengue virus infection in Africa. *Emerg Infect Dis* 2011;**17**(8):1349–54.

36. Ranjit S, Kissoon N. Dengue hemorrhagic fever and shock syndromes. *Pediatr Crit Care Med* 2011;**12**(1):90–100.

37. Bhatt S, Gething PW, Brady O, et al. The global distribution and burden of dengue. *Nature* 2013;**496**(7446):504–7.

38. Carod-Artal F, Wichmann O, Farrar J, Gascón J. Neurological complications of dengue virus infection. *Lancet Neurol* 2013;**12**(9):906–19.

39. Chen L, Wilson M. Dengue and chikungunya infections in travelers. *Curr Opin Infect Dis* 2010;**23**(5):438–44.
40. Global strategy for dengue prevention and control (PDF). World Health Organization, 2012, pp. 16–17.
41. Comprehensive guidelines for prevention and control of dengue and dengue haemorrhagic fever. (PDF) (Rev. and expanded. ed.). New Delhi, India: World Health Organization Regional Office for South-East Asia, 2011, p. 17.
42. Gubler D. Dengue and dengue hemorrhagic fever. *Clin Microbiol Rev* 1998;**11**(3):480–96.
43. Guzman M, Halstead S, Artsob H, et al. Dengue: a continuing global threat. *Nat Rev Microbiol* 2010;**8**(12 Suppl):S7–16.
44. Halstead SB. *Dengue*. London: Imperial College Press; 2008. p. 180 & 429.
45. Kularatne S. Dengue fever. *BMJ* 2015;**351**:h4661.
46. Martina B, Koraka P, Osterhaus A. Dengue virus pathogenesis: an integrated view. *Clin Microbiol Rev* 2009;**22**(4):564–81.
47. Paixão E, Teixeira M, Costa M, Rodrigues LC. Dengue during pregnancy and adverse fetal outcomes: a systematic review and meta-analysis. *Lancet Infect Dis* 2016;**16**(7):857–65.
48. Rodenhuis-Zybert I, Wilschut J, Smit JM. Dengue virus life cycle: viral and host factors modulating infectivity. *Cell Mol Life Sci* 2010;**67**(16):2773–86.
49. Simmons C, Farrar J, Nguyen V, Wills B. Dengue. *N Engl J Med* 2012;**366**(15):1423–32. pp. 25–27.
50. Varatharaj A. Encephalitis in the clinical spectrum of dengue infection. *Neurol India* 2010;**58**(4):585–91.
51. Webster D, Farrar J, Rowland-Jones S. Progress towards a dengue vaccine. *Lancet Infect Dis* 2009;**9**(11):678–87.
52. WHO. Chapter 2: clinical diagnosis. Dengue haemorrhagic fever: diagnosis, treatment, prevention and control(PDF) (2nd ed.). Geneva: World Health Organization, 1997, pp. 12–23.
53. Wilder-Smith A, Chen L, Massad E, Wilson M. Threat of dengue to blood safety in dengue-endemic countries. *Emerg Infect Dis* 2009;**15**(1):8–11.
54. Whitehorn J, Farrar J. Dengue. *Br Med Bull* 2010;**95**:161–73.
55. Wiwanitkit V. Dengue fever: diagnosis and treatment. *Expert Rev Anti Infect Ther* 2010;**8**(7):841–5.
56. Yacoub S, Wills B. Predicting outcome from dengue. *BMC Med* 2014;**12**(1):147.
57. Zheng Y. *Mosquito-borne virus infects 2d in Mass*. Boston, MA: The Boston Globe; 2008.
58. Eastern equine encephalitis. CDC, 2010. Retrieved August 7, 2012.
59. Eastern equine encephalitis fact sheet. CDC, 2010. Retrieved August 30, 2015.
60. Campbell G, Hills S, Fischer M, Jacobson J, Hoke C, Hombach J, et al. Estimated global incidence of Japanese encephalitis: a systematic review. *Bull World Health Organ* 2011;**89**(10):766–74.
61. Ghosh D, Basu A. Japanese encephalitis-a pathological and clinical perspective. In: Brooker S, editor. *PLoS Negl Trop Dis* 2009;**3**(9):e437.
62. Kim HC, Terry AK, RatreeTakhampunya B, Evans P, SirimaMingmongkolchai AK, Grieco J, et al. Japanese Encephalitis virus in Culicine mosquitoes (Diptera: Culicidae) collected at Daeseongdong, a village in the Demilitarized Zone of the Republic of Korea. *J Med Entomol* 2011;**48**(6):1250–6.
63. Ghoshal A, Das S, Ghosh S, Mishra MK, Sharma V, Koli P, et al. Proinflammatory mediators released by activated microglia induces neuronal death in Japanese encephalitis. *Glia* 2007;**55**(5):483–96.
64. Jelinek T. Japanese encephalitis vaccine in travelers. *Expert Rev Vaccines* 2008;**7**(5):689–93.

65. Gupta N, Lomash V, LakshmanaRao PV. Expression profile of Japanese encephalitis virus induced neuroinflammation and its implication in disease severity. *J Clin Virol* 2010;**49**(1):04–10.

66. Solomon T, Dung N, Kneen R, Gainsborough M, Vaughn D, Khanh V. Japanese encephalitis. *J Neurol Neurosurg Psychiatry* 2000;**68**(9):405–15.

67. Su H, Liao C, Lin Y. Japanese encephalitis virus infection initiates endoplasmic reticulum stress and an unfolded protein response. *J Virol* 2002;**76**(9):4162–71.

68. Schiøler K, Samuel M, Wai K. Vaccines for preventing Japanese encephalitis. *Cochrane Database Syst Rev* 2007;**3**:CD004263.

69. MedTerms™ Medical Dictionary: http://www.medicinenet.com/script/main/art.asp?articlekey= 38474.

70. Murray valley encephalitis (MVE) factsheet. New South Wales Department of Health, 2011. Retrieved 30 December 2011.

71. Burnet F. Murray valley encephalitis. *Am J Public Health Nations Health* 1952;**42**(12):1519–21.

72. French E. Murray valley encephalitis isolation and characterization of the aetiological agent. *Med J Aust* 1952;**1**(4):100–3.

73. Hurrelbrink R, Nestorowicz A, McMinn P. Characterization of infectious Murray valley encephalitis virus derived from a stably cloned genome-length cDNA. *J Gen Virol* 1999;**80**(12):3115–25.

74. Thompson W, Kalfayan B, Anslow RO. Isolation of California encephalitis virus from a fatal human illness. *Am J Epidemiol* 1965;**81**:245–53.

75. La Crosse virus disease cases and deaths reported to CDC by year and clinical presentation, 2004–2013 (PDF). Centers for Disease Control and Prevention. Retrieved 4 December 2016.

76. Center for Disease Control and Prevention (CDC). Possible congenital infection with La Crosse Encephalitis virus—West Virginia, 2006-2007. *MMWR Morb Mortal Wkly Rep* 2009;**58**(1):4–7.

77. McJunkin J. Safety and pharmacokinetics of ribavirin for the treatment of la crosse encephalitis. *Pediatr Infect Dis J* 2011;**30**(10):860–5.

78. McJunkin J, de los Reyes E, Irazuzta J, Caceres M, Khan R, Minnich L, et al. La Crosse Encephalitis in Children (pdf). *N Engl J Med* 2001;**344**(11):801–7.

79. Bhatt S, Weiss D, Cameron E, Bisanzio D, Mappin B, Dalrymple U, et al. The effect of malaria control on Plasmodium falciparum in Africa between 2000 and 2015. *Nature* 2015;**526**(7572):207–11.

80. Korenromp E, Williams B, de Vlas S, Gouws E, Gilks C, Ghys P, et al. Malaria attributable to the HIV-1 epidemic, sub-Saharan Africa. *Emerg Infect Dis* 2005;**11**(9):1410–9.

81. Layne, S. Principles of Infectious Disease Epidemiology(PDF). EPI 220. UCLA Department of Epidemiology. Retrieved 15 June 2007.

82. Murray C, Rosenfeld L, Lim S, Andrews K, Foreman K, Haring D, et al. Global malaria mortality between 1980 and 2010: a systematic analysis. *Lancet* 2012;**379**(9814):413–31.

83. Olupot-Olupot P, Maitland K. Management of severe malaria: results from recent trials. *Adv Exp Med Biol* 2013;**764**:241–50.

84. World Malaria Report 2015. World Health Organization, 2015.

85. Cowman A, Berry D, Baum J. The cellular and molecular basis for malaria parasite invasion of the human red blood cell. *J Cell Biol* 2012;**198**(6):961–71.

86. Tilley L, Dixon M, Kirk K. The Plasmodium falciparum-infected red blood cell. *Int J Biochem Cell Biol* 2011;**43**(6):839–42.

87. Tran T, Samal B, Kirkness E, Crompton P. Systems immunology of human malaria. *Trends Parasitol* 2012;**28**(6):248–57.

88. Vaughan A, Aly A, Kappe S. Malaria parasite pre-erythrocytic stage infection: gliding and hiding. *Cell Host Microbe* 2008;**4**(3):209–18.

89. Beare N, Taylor T, Harding S, Lewallen S, Molyneux M. Malarial retinopathy: a newly established diagnostic sign in severe malaria. *Am J Trop Med Hyg* 2006;**75**(5):790–7.

90. Greenwood B, Bojang K, Whitty C, Targett G. Malaria. *Lancet* 2005;**365**(9469):1487–98.

91. Koella J, Sorensen A. The malaria parasite, Plasmodium falciparum, increases the frequency of multiple feeding of its mosquito vector, Anopheles gambiae. *Proc R Soc B* 1998;**265**(1398):763–8.

92. Lalloo D, Olukoya P, Olliaro P. Malaria in adolescence: burden of disease, consequences, and opportunities for intervention. *Lancet Infect Dis* 2006;**6**(12):780–93.

93. Nadjm B, Behrens R. Malaria: an update for physicians. *Infect Dis Clin North Am* 2012;**26**(2):243–59.

94. Organization, World Health. *Guidelines for the treatment of malaria.* 2nd ed. Geneva: World Health Organization; 2010.

95. White N. Determinants of relapse periodicity in Plasmodiumvivax malaria. *Malar J* 2011;**10**:297.

96. WHO. *World Malaria Report 2014.* Geneva, Switzerland: World Health Organization; 2014 . pp. 32–42.

97. Baird J. Evidence and implications of mortality associated with acute Plasmodium vivax malaria. *Clin Microbiol Rev* 2013;**26**(1):36–57.

98. Idro R, Marsh K, John C, Newton C. Cerebral malaria: mechanisms of brain injury and strategies for improved neurocognitive outcome. *Pediatr Res* 2010;**68**(4):267–74.

99. Kochar D, Saxena V, Singh KN, Kumar S, Das A. Plasmodium vivax malaria. *Emerg Infect Dis* 2005;**11**(1):132–4.

100. Rénia L, Wu Howland S, Claser C, Charlotte Gruner A, Suwanarusk R, HuiTeo T, et al. Cerebral malaria: mysteries at the blood-brain barrier. *Virulence* 2012;**3**(2):193–201.

101. Sarkar P, Ahluwalia G, Vijayan V, Talwar A. Critical care aspects of malaria. *J Intensive Care Med* 2009;**25**(2):93–103.

102. Trampuz A, Jereb M, Muzlovic I, Prabhu R. Clinical review: severe malaria. *Crit Care* 2003;**7**(4):315–23.

103. Achan J, Talisuna A, Erhart A, Yeka A, Tibenderana J, Baliraine F, et al. Quinine, an old antimalarial drug in a modern world: role in the treatment of malaria. *Malar J* 2011;**10**(1):144.

104. Kajfasz P. Malaria prevention. *Int Marit Health* 2009;**60**(1–2):67–70.

105. Kokwaro G. Ongoing challenges in the management of malaria. *Malar J* 2009;**8**(Suppl 1):S2.

106. Waters N, Edstein M. 8-Aminoquinolines: Primaquine and tafenoquine. In: Staines H, Krishna S, editors. *Treatment and prevention of malaria: antimalarial drug chemistry action and use.* Basel: Springer; 2012. p. 69–93.

107. Pasvol G. The treatment of complicated and severe malaria. *Br Med Bull* 2005;**75–76**:29–47.

108. Sinha S, Medhi B, Sehgal R. Challenges of drug-resistant malaria. *Parasite* 2014;**21**:61.

109. Auguste A, Pybus O, Carrington C. Evolution and dispersal of St. Louis encephalitis virus in the Americas. *Infect Genet Evol* 2009;**9**(4):709–15. .Baillie G, Kolokotronis S, Waltari E, Maffei J, Kramer L, Perkins S. Phylogenetic and evolutionary analyses of St. Louis encephalitis virus genomes. *Mol Phylogenet Evol* 2008;**47**(2):717–28.

110. Bredeck J. The story of the epidemic of Encephalitis in St Louis. *Am J Public Health Nations Health* 1933;**23**(11):1135–40.

111. Kramer L, Chandler L. Phylogenetic analysis of the envelope gene of St. Louis encephalitis virus. *Arch Virol* 2001;**146**(12):2341–55.

112. May F, Li L, Zhang S, Guzman H, Beasley D, Tesh R, et al. Genetic variation of St. Louis encephalitis virus. *J Gen Virol* 2008;**89**(8):1901–10.

113. Encephalitis in St. Louis. Am J Public Health Nation's Health 1993;23(10): 1058–1060.

114. Rahal J, Anderson J, Rosenberg C, Reagan T, Thompson L. Effect of interferon-alpha2b therapy on St. Louis viral meningoencephalitis: clinical and laboratory results of a pilot study. *J Infect Dis* 2004;**190**(6):1084–7.

115. Barrett A, Higgs S. Yellow fever: a disease that has yet to be conquered. *Annu Rev Entomol* 2007;**52**:209–29.

116. Auguste A, Lemey P, Pybus O, Suchard M, Salas R, Adesiyun A, et al. Yellow fever virus maintenance in Trinidad and its dispersal throughout the Americas. *J Virol* 2010;**84**(19):9967–77.

117. Barrett A, Higgs S, Yellow fever: a disease that has yet to be conquered. Annu Rev Entomol 2007;52: 209–229. Communicable diseases manual, 19th edition. Washington, DC: American Public Health Association.

118. Frequently asked questions about yellow fever. CDC, 2015. Retrieved 18 March 2016.

119. Rogers D, Wilson A, Hay S, Graham A. The global distribution of yellow fever and dengue. *Adv Parasitol* 2006;**62**:181–220.

120. Staples J, Monath T. Yellow fever: 100 years of discovery. *JAMA* 2008;**300**(8):960–2.

121. Australian Government. The Australian Immunisation Handbook, 10th ed., 2013.

122. Palmer SR. *Oxford textbook of zoonoses: biology, clinical practice, and public health control*. 2nd ed. Oxford: Oxford University Press; 2011. p. 423.

123. Arzt J, White WR, Thomsen BV, Brown C. Agricultural diseases on the move early in the third millennium. *Vet Pathol* 2010;**47**(1):15–27.

124. Bird B, Ksiazek T, Nichol S, Maclachlan N. Rift Valley fever virus. *J Am Vet Med Assoc* 2009;**234**(7):883–93.

125. WHO, Rift Valley fever, Fact sheet N207, 2014. May 2010. Retrieved 21 March 2014.

126. Rift Valley fever. Fact sheet N207. World Health Organization, 2010. Retrieved 21 March 2014.

127. Rift Valley Fever reviewed and published by WikiVet, Accessed 12 October 2011.

128. Boshra H, Lorenzo G, Busquets N, Brun A. Rift valley fever: recent insights into pathogenesis and prevention. *J Virol* 2011;**85**(13):6098–105.

129. Swanepoel R, Coetzer J. Rift Valley fever. In: Coetzer J, Tustin RC, editors. *Infectious diseases of livestock*. 2nd ed. CapeTown: OxfordUniversityPressSouthernAfrica; 2004. p. 1037–70.

130. Bird BH, Maartens LH, Campbell S, Erasmus JB, Erickson BR, Dodd KA, et al. Rift valley fever virus vaccine lacking the NSs and NSm genes is safe, nonteratogenic, and confers protection from viremia, pyrexia, and abortion following challenge in adult and pregnant Sheep ▽. J Virol 2011;**85**(24):12901–9.

131. Hirsh CD, Maclachlan NJ, Walker RL. *Veterinary microbiology*. Hoboken, NJ: Wiley-Blackwell; 2004. p. 354.

132. Cook CG, Zumla AI. *Manson's tropical diseases*. Philadelphia, PA: Elsevier Health Sciences; 2008. pp. 736–737.

133. Mackenzie JS, Ashford R, Service MW. *Encyclopedia of arthropod-transmitted infections of man and domesticated animals*. Oxfordshire: CABI; 2001. p. 251.

134. Department of Health And Ageing—Kunjin virus infection—Fact Sheet. Government of Australia, 2004. Retrieved 8 August 2009.

135. Cook G,C, Zumla AI. *Manson'stropicaldiseases*. Philadelphia, PA: ElsevierHealthSciences; 2008. pp. 736–737.

136. Harley D, Sleigh A, Ritchie S. Ross River virus transmission, infection, and disease: a cross-disciplinary review. *Clin Microbiol Rev* 2001;**14**(4):909–32. table of contents.

137. Fraser J. Epidemic polyarthritis and Ross River virus disease. *Clin Rheum Dis* 1986;**12**:369–88.
138. Ross river virus disease. Better Health Channel. Department of Health, State Government, Victoria, 2013. Retrieved 15 April 2015.
139. Schleenvoigt B, Baier M, Hagel S, Forstner C, Kotsche R, Pletz M. Ross river virus infection in a Thuringian traveller returning from south-east Australia. *Infection* 2015;**43**(2):229–30.
140. Cashman P, Barmah Forest virus serology: implications for diagnosis and public health action. Communicable Diseases Intelligence, 2008. Retrieved 7 April 2015.
141. Kostyuchenko V, Jakana J, Liu X, Haddow A, Aung M, Weaver S, et al. The structure of Barmah forest virus as revealed by cryo-electron microscopy at a 6-angstrom resolution has detailed transmembrane protein architecture and interactions. *J Virol* 2011;**85**(18):9327–33.
142. Smith D. The viruses of Australia and the risk to tourists. *Travel Med Infect Dis* 2011;**9**:113–25.
143. Ehlkes L. Surveillance should be strengthened to improve epidemiological understandings of mosquito-borne Barmah Forest virus infection. *West Pacific Surveill Response J* 2012;**3**:63–8.
144. Naish S, Mengersen K, Hu W, Tong S. Forecasting the future risk of Barmah forest virus disease under climate change scenarios in Queensland, Australia. *PLoS One* 2013;**8**(5):E62843.
145. Naish S, Hu W, Tong S. Spatio-temporal patterns of Barmah forest virus disease in Queensland, Australia. *PLoS One* 2011;**6**(10):e25688.
146. Pastula D, Hoang Johnson D, White J, Dupuis 2nd A, Fischer M, Staples J. Jamestown Canyon virus disease in the United States—2000–2013. *Am J Trop Med Hyg* 2015;**93**(2):384–9.
147. Lindsey N, Lehman J, Staples J, Fischer M, Division of Vector-Borne Diseases, National Center for Emerging and Zoonotic Infectious Diseases, CDC. West Nile virus and other arboviral diseases—United States, 2013. *MMWR Morb Mortal Wkly Rep* 2014;**63**(24):521–6. CDC.

FURTHER READING

148. Arrigo N, Adams A, Weaver S. Evolutionary patterns of eastern equine encephalitis virus in North versus South America suggest ecological differences and taxonomic revision. *J Virol* 2010;**84**(2):1014–25.
149. Eastern equine encephalitis. CDC, 2010. Retrieved 7 August 2012.
150. Eastern equine encephalitis fact sheet. CDC, 2010. Retrieved 30 August 2015.

Chapter 3

Flaviviruses

Chapter Outline

Flaviviruses continue to fascinate epidemiologists, immunologists, virologists, entomologists, and the health-care providers from Africas to Americas. These are maintained in nature by complex transmission cycles. Research over the past few decades has unraveled many aspects of flaviviral behavior in natural conditions and in the laboratory yet the recent epidemics caused by these viruses in South America point toward a multifaceted virulence mechanisms. Flaviviruses cause a variety of human diseases ranging from mild febrile illnesses to severe hemorrhagic manifestations. The purpose of this review is to summarize what is presently known about the history, taxonomy, and the structure of these viruses. In the later part of this chapter, we will focus on the transmission cycles of Zika virus and the prevention strategies against flaviviruses.

THE OUT OF AFRICA PHENOMENON

In the summer of 1878, Kate Bionda in Memphis Tennessee died of "a yellow plague" after a steamboat worker, William Warren, escaped a quarantined steamboat and visited her restaurant for a lunch.[1] It was only the beginning of a devastating epidemic that killed nearly 15,000 people and forced 20,000 people to leave the city.[2] The infected population had spiking fevers, failing organs, and their skin and eyes turned yellow. This epidemic forced 20,000 residents to evacuated to adjoining states within a week of eruption, and by the end of epidemic the city was left with less than half of its original population. Many of the fleeing citizens were under "shotgun quarantine" when they arrived other towns and often they were made to wear yellow jackets as a means of identification.[3]

Zika Virus Disease. https://doi.org/10.1016/B978-0-12-812365-2.00004-4

Many activists tried to get the establishment funding for quarantine facilities, but were turned down as very little was known about the mode of its transmission. An average of 200 people died every day through September of 1878. There were corpses everywhere and near continual ringing of funeral bells. Half of the city's doctors died in the outbreak.[1]

The disease historians in the United States date the initial cases in Africa and arrival of the disease in North America back to the 17th century.[4,5] The disease is believed to be imported by slave ships from the Caribbean. Before 1822, small epidemics have been reported as far north as Boston, but since the later part of 18th century the disease has been restricted to the Southern costal towns. The disease eventually spread up the Mississippi river system rampaging New Orleans, Mobile, Savannah, Charleston, and Memphis (Fig. 3.1).[6]

At that time, germ theory was still a newer concept. The doctors had no idea that the cause of this sickness is a Yellow Fever virus, which is thousands of times smaller than the naked eye can see, and that the mosquitoes were the carriers. It was not until 1927 that researchers isolated the culprit virus. Since it caused the patients to turn yellow and would also become the representative virus of the group of similar viruses, the family will take its name from the French word "flavum" meaning "yellow." Interestingly, the first mammalian viruses to be identified included four arthropod vector-borne viruses, three of which were Flaviviruses: Louping Ill virus, Yellow Fever virus, and Dengue virus (Fig. 3.2).

The Flaviviruses constitute a fascinating group of viruses that exhibit clear correlations between phylogenetic relationships and virus-vector-host interactions.[7] These viruses are the causative agents of significant morbidity and mortality among humans and domestic animals.

TAXONOMY OF FLAVIVIRUSES

The viruses in the Flavivirus family are all single-stranded ribonucleic acid viruses. Their genomes are arranged in a linear nonsegmented configuration. As their proteins are made directly from the open-frame, and template strand of ribonucleic acid are present in the viral capsid, they are considered positive sense (+). The genome of these viruses is on average 11,000 nucelotides long and encodes around 10 genes.[8]

The capsid (C) protein is icosahedral shaped and enclosed in a spherical envelope. It is highly basic protein of ~11 kDa and approximately 40–50 nm in diameter (Fig. 3.3). Charged residues are clustered at the N- and C-termini, separated by an internal hydrophobic region that mediates membrane association.[9,10] The amateur C-protein has a c-terminal hydrophobic anchor that serves as a signal peptide for endoplasmic reticulum translocation of prM-protein. This hydrophobic domain is ultimately cleaved from mature C-protein by the viral serine protease. Mature C-protein folds into a compact dimer with each monomer containing four alpha helices.

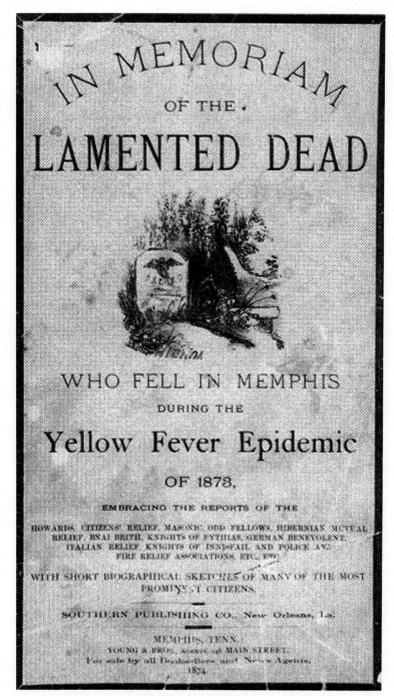

FIG. 3.1 Cover of the book "In Memoriam of the Lamenated Dead." The book accounts short biographical sketches of many prominent citizens who lost their lives during the yellow fever epidemic 1878 in Memphis, Tennessee.

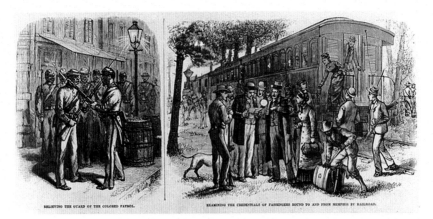

FIG. 3.2 The exodus from Memphis during Yellow Fever epidemic of 1878 and shotgun quarantine.

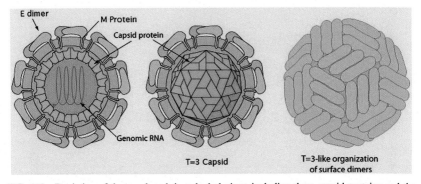

FIG. 3.3 Depiction of the enveloped, icosahedral virus, including three capsid proteins and the genetic material of Flaviviruses. *Picture source: http://www.expasy.org.*

The surface of viral particles contains two viral proteins, E (envelope) and M (membrane).[11] The E glycoprotein, the major antigenic determinant on virus particles, contains a cellular receptor-binding site(s) and a fusion peptide.[12,13] It mediates binding and fusion during virus entry. E-protein is synthesized as a type I membrane protein consisting of 12 conserved cysteines that form disulfide bonds. Proper folding, stabilization in low pH, and secretion of E depend on coexpression with prM-protein.[14–16]

The M protein, produced during maturation of nascent virus particles within the secretory pathway, is a small proteolytic fragment of the membrane glycoprotein prM-protein.[17] Before the prM-protein is cleaved to yield pr peptide and M protein, it might function as a chaperone for folding and assembly of E-protein.[18] prM is translocated into the endoplasmic reticulum by the C-terminal hydrophobic domain of C-protein (Table 3.1).

TABLE 3.1 Members of genus Flavivirus

Taxonomic unit	Representative examples
Tick-borne viruses	
Mammalian tick-borne group	Tick-borne encephalitis virus, European subtype
	Tick-borne encephalitis virus, Far Eastern subtype
Seabird tick-borne group	Saumarez Reef virus
	Tyuleniy virus
Mosquito-borne viruses	
Aroa virus group	Aroa group
Dengue virus group	Dengue virus, types 1 to 4
	Kedougou virus
Japanese encephalitis group	Japanese encephalitis virus
	West Nile virus
Kokobera virus group	Kokobera virus
Ntaya virus group	Ntaya virus
Spondweni virus group	Spondweni virus
Yellow fever virus group	Yellow fever virus
Viruses with no known vector	
Entebbe bat virus group	Entebbe bat virus
Modoc virus group	Modoc virus
Rio Bravo virus group	Rio Bravo virus
Unclassified	Cell fusing agent virus

STRUCTURE

The single, positive-strand RNA of Flaviviruses has a 5′ type I cap that differentiates them from viruses of the other genera. Additional methylation of N2 residue constitutes type II cap and has been detected in the ribonucleic acid from defected cells. However, unlike cellular messenger ribonucleic acid, the genome of Flavivirus lacks a 3′ poly-A tail, but forms a loop structure. The sequence of the 5′ noncoding regions is not well conserved between different flaviviruses, although common secondary structures have been found within this region. These structures are involved in the translation of genome. The organization of 3′ noncoding regions differs greatly between mosquito-borne and tick-borne viruses, but despite this variability similar patterns of conserved sequences have been found among Falviviruses.[19]

In Flaviviruses, structural genes are found at the 5′ of the genome and non-structural genes are encoded at the 3′ end of the genome.[20] This organization allows the virus to maximize the production of structural genes, since viral assembly requires more structural genes to be made than nonstructural proteins.[21,22]

REPLICATION

The Flavivirus replicase associates with membranes through interactions involving the small hydrophobic nonstructural proteins, viral ribonucleic acid, and presumably some host factors.[23] Flaviviruses enter host cells by receptor-mediated endocytosis.[24,25] Replication of Flaviviruses takes place in the cytoplasm. They cannot replicate in the nucleus because it uses host cell's RNA-dependent RNA polymerase to replicate. The trimerization of viral E-protein, triggered by acidic environment of endosome, enables fusion of viral and cellular membranes. Subsequently, the neucleocapsid is released into the cytoplasm and capsid protein and ribonucleic acid dissociate. The replication of ribonucleic acid genome and particle assembly is initiated in the lumen of endoplasmic reticulum. Replication begins with the synthesis of minus strand ribonucleic acid, which then serves as a template for the synthesis of plus strand ribonucleic acid. Viral ribonucleic acid replicates in an asymmetric manner producing tenfold excess of plus strands over minus strands.[26] Three major species of labeled Flavivirus ribonucleic acid have been described, including genome ribonucleic acid, a double-stranded replicative form, and a heterogeneous population of replicative intermediate RNA. The mode of Flavivirus replication can thus be confidently described as semiconservative and asymmetric.

The process of viral replication triggers a series of ultrastructural changes in the Flavivirus infected cells. The earliest event is proliferation of endoplasmic reticulum followed by production of smooth membrane structures corresponding to the exponential viral replication. These vesicles are 70- to 200-nm vesicles within the lumen of smooth endoplasmic reticulum. Viral ribonucleic acid accumulates in association with these vesicle packets in the perinuclear region of mammalian cells.

The efficiency of genome translation can be a primary determinant of Flavivirus infectivity. Therefore, Flaviviruses use several mechanisms to facilitate translational competence, including specialized structures within the 5′ and 3′ noncoding regions. Translational is cap dependent and initiates by ribosomal scanning. Translation of the single, long open reading frame produces a large polyprotein that is posttranslationally processed into 10 proteins. The sequences closer to the N-terminal encode structural proteins (C-prM-E), followed by the nonstructural proteins (NS1-NS2A-NS2B-NS3-NS4A-2K-NS4B-NS5). Host signal peptidase and virus-encoded serine protease cleave the structural and nonstructural proteins.[27]

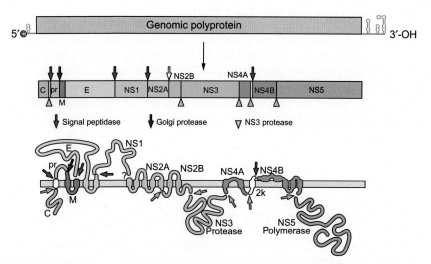

FIG. 3.4   Flavivirus genome organization and polyprotein processing. The virion RNA is about 11 kb. At the top is the viral genome with the structural and nonstructural protein coding regions and the 5'- and 3'-NCRs. Boxes below the genome indicate viral proteins generated by the proteolytic processing cascade.

The immature particles that are produced initially are noninfective and undergo subsequent cleavage in Golgi body network to produce mature, infectious particles. Mature virus and subviral particles are released from the host cell by exocytosis.[28] Most Flaviviruses are transmitted to vertebrate hosts by arthropod vectors, mosquitoes, or ticks, in which they replicate actively. Some Flaviviruses transmit between rodents or bats without known arthropod vectors (Fig. 3.4).

EVOLUTION

Flaviviruses constitute a diverse group of viruses with over 85 known species. They cause a variety of diseases, including fevers, encephalitis, and hemorrhagic fevers. The entities of human public health significance include Dengue Hemorrhagic fever, Yellow Fever, and Zika viruses. All of these have two ecologically and evolutionarily distinct transmission cycles:

- A sylvatic transmission cycle—Virus circulates between zoonotic vertebrate reservoir and amplification hosts and arboreal mosquitoes
- An urban transmission cycle—Virus circulates between humans and peridomestic *Aedes* spp. mosquitoes

Due to remarkable similarities between Zika virus and other mosquito-borne viruses of Flavivirus family, we will focus on the transmission patterns of Zika virus in the following sections.

INTRODUCTION OF THE ZIKA VIRUS

Since it was first identified in rhesus monkeys in Zika Forest preserve in Uganda, about half a century ago, Zika virus did not get substantial attention because of its rarity and potential confinement to a few African countries.[29] Later the virus was isolated from *Aedes africanus* mosquitoes and the first human cases with Zika virus infection were reported in Nigeria in 1952. Only sporadic human cases were reported in the following 60 years in Malaysia and Indonesia.[30] In 2007, Zika virus infection out broke in Yap Island of Micronesia in 2007, followed by an epidemic in French Polynesia in 2013. As of April 2016, Zika virus infected cases have been reported in 35 countries in the Americas, predominantly South America where it has reached pandemic levels.

TRANSMISSION CYCLES OF ZIKA VIRUS

As mentioned earlier, the Zika virus circulation has been documented in two ecologically and evolutionarily distinct transmission cycles: a sylvatic transmission cycle and an urban transmission cycle.

The initial documented isolations of Zika virus was from *Aedes (stegomyia) africanus* mosquitoes in Uganda, Central African Republic, and Senegal.[31,32] Subsequently, the virus has also been identified in *Aedes (stegomyia) luteocephalus*, *Aedes (stergomyia) opok*, *Aedes (stegomyia) apicoargenteus*, *Aedes (stegomyia) luteocephalus*, and many other Aedene and non-Aedene mosquito species.[33] The isolation of Zika virus from *Aedes (Fredwardsius) vittatus* sampled within the "zone of emergence" indicates its potential role as a bridge vector for introducing enzootic Zika virus strains into the human transmission cycle.

In the urban life cycle, the Zika virus transmission mainly involves *Aedes (stegomyia) aegypti* mosquito.[34] Other Aedian species such as *Aedes (stegomyia) hensilli*, *Aedes (Stegomyia) hensilli*, *Aedes (Stegomyia) polynesiensis*, and *Aedes (Stegomyia) albopictus* have also been identified as potential urban vectors.

The recent striking spread of Zika virus has been attributed to its adaptive evolution in Southeast Asia or the South Pacific for more efficient urban transmission by *Aedes aegypti* mosquito. Also the immunologically naïve human populations in the Americas have also contributed to the extent and intensity of its transmission.

Five distinct modes of nonvector transmission in humans have been identified:

- Sexual transmission—Zika virus has been identified in semen samples up to 62 days after the onset of illness. The Center for Disease Control and Prevention has reported two confirmed and four suspected cases of sexual transmission from symptomatic male partners with recent travel history to Central America to their female partners with no history of travel outside continental United States of America.[35,36] Furthermore, the first case of male-to-male transmission has also been reported from Texas.[37] Zika virus is unique in its mode of sexual transmission, since none of the other mosquito-borne arboviruses are transmitted by this route.

- Vertical transmission—Two cases from French Polynesia report vertical transmission of Zika virus from mothers to their infants.[38,39] The virus was detected in the serum of both mothers and their infants and peripartum transmission were suspected. Furthermore, Zika virus has been demonstrated by reverse transcription polymerase chain reaction in the brain tissue of many neonates with microcephaly whose mothers had reported rashes during first trimester.[40,41] There is still debate regarding the period of pregnancy during which the maternal infection incurs developmental abnormalities to the fetus.[40]

- Breastfeeding—After the outbreak in French Polynesia, Zika virus was detected in the breast milk and saliva of infected mothers, suggesting a potential transmission route.[42,43] However, the transmissibility and the ultimate clinical presentation of an infant infected by breast milk has not been established and further studies on the presence of viral RNA and virus reactivity in breast milk are required.

- Blood transfusion—During the outbreak in French Polynesia, 2.8% of the asymptomatic blood donors tested positive for Zika virus by PCR.[44] This transmission route can introduce novel agents into the susceptible population especially the immunocompromised and elderly. Also, the scale of transfusion-transmitted infection may have huge public heath impact especially for the economically challenged health-care sectors.[45]

- Occupational exposure—A case of needle stick injury associated with Zika virus transmission has been reported from personnel of a United States laboratory.

VACCINES FOR FLAVIVIRUSES

The development of the first live-attenuated Flavivirus vaccine, yellow fever virus strain 17D led to Max Theiler's Nobel Prize for Physiology or Medicine in 1951. Only a limited number of Flavivirus vaccines are available, including inactivated tick-borne encephalitis virus and Japanese encephalitis virus for use in humans and inactivated West Nile virus for use in animals.[46] Development of effective Dengue virus vaccines that exhibit cross-protection, thought to be important for preventing subsequent Dengue virus-associated immunopathogenesis are particularly challenging. The major virion surface protein is E, which is involved in receptor binding and membrane fusion. Therefore, most neutralizing antibodies are induced against E-protein, and all approved (and most developing) Flavivirus vaccines contain E antigens. Effective Flavivirus vaccines against Yellow Fever virus, Japanese encephalitis virus, and Tick-borne encephalitis virus infections have been developed, but West Nile virus and Dengue virus vaccines have not been licensed for human use.[47] The ability to genetically manipulate Flaviviruses is being used to develop novel approaches, including live-attenuated chimeric vaccines to other Flaviviruses based on the Yellow Fever virus-17D vaccine template.

Due to its links with debilitating birth defects, Zika virus has set-off world-wide alarm and the drug companies are engaged in an all-out race to develop a Zika virus vaccine. Usually vaccines may take a decade or more to develop, but researchers believe that they will be able to crack the virus code by as early as 2018. The ambitious vaccine developers are trying the traditional dead-virus vaccine method and also the innovative technologies that rely on splicing deoxyribonucleic acid. Many clinical trial organizers are all set to begin their initial testing for vaccines in South America by 2017 at the height of summer.

CHALLENGES TO VECTOR CONTROL

Due to the absence of licensed vaccines and therapeutics, the only workable methods of reducing Zika virus spread are reduction in mosquito populations and human-vector contact. Rapid urbanization and relative reluctance in use of environmentally unacceptable pesticides like dichlorodiphenyltrichloroethane have weakened the vector control measures and the tropical cities have been re-infested with *Aedes aegypti*. In current circumstances, *Aedes aegypti* population can be controlled by the following cost-effective measures:

- Spreading public awareness for eliminating larval habitats such as standing water in flower pots and water storage containers
- Use of insecticide aerosols where people are exposed to vectors
- Release of genetically modified mosquitoes that express a lethal gene causing death of all offspring from mating with wild females
- Use of mosquito traps

VIRAL RESISTANCE TO HOST DEFENSES

Positive-stranded ribonucleic acid viruses are capable of evolving and evading the antiviral defense mechanisms such as interferons. Flaviviral nonstructural proteins inhibit interferon-induced signaling through interferon receptor and JAK-STAT pathway. These viruses are also capable of controlling the intrinsic cellular antiviral response pathways such as preventing the induction of interferons by avoiding the early activation of interferon regulatory factor 3.

PERSPECTIVES

Although the research on Flaviviruses is dynamic and is adding quickly to our knowledge of these pathogens, large gaps exist in our understanding of every step in their complex life cycle. The improved genetic and biochemical tools have added substantially to identifying host cell surface molecules that could be involved in viral binding and entry. In-depth studies on the viral transcription, translation, and polyprotein processing events are needed to identify potential targets of drug therapy. Newer information has emerged on the accelerating features of nonstructural viral proteins for genome replication, which may become potential targets for virus eradication. Our knowledge regarding virion formation and egress is still at an early age and it is not clear how structural proteins combine to form nascent virions and where the process is regulated. Clearly answers to all these questions will allow us to address the unique aspects of this distinct viral family in order to develop effective immunization and therapeutic strategies.

FLAVIVIRUSES AND RIO OLYMPICS-2016

Among many controversies that surrounded Rio Olympics-2016, the prevailing threat was from Flaviviruses since Brazil has been the nexus of Zika virus infection outbreak. Professor Amir Attaran, Associate Professor of Law and Population Health published a report in *Harvard Public Health Review* suggesting that 2016 Olympics could cause "full-blown public health disaster." The famous five key points from the report are as below:

> *First, Rio de Janeiro is more affected by Zika than anyone expected, rendering earlier assumptions of safety obsolete....Second, although Zika virus was discovered nearly seventy years ago, the viral strain that recently entered Brazil is clearly new, different, and vastly more dangerous than "old" Zika....Third, while Brazil's Zika inevitably will spread globally—given enough time, viruses always do—it helps nobody to speed that up....Fourth, when (not if) the Games speed up Zika's spread, the already-urgent job of inventing new technologies to stop it becomes harder....Fifth, proceeding with the Games violates what the Olympics stand for. The International Olympic Committee writes that "Olympism seeks to create... social responsibility and respect for universal fundamental ethical principles." But how socially responsible or ethical is it to spread disease?[48]*

The tourists and athletes from around the world were not only preparing for competitive sport but were also reading terrible facts about Zika virus disease, such as its links to debilitating neurological conditions and birth defects.[49] Jessica Ennis-Hill, a British track star Minichiello's coach, told the Daily mail: "Jessica very much wants to have more children so we are taking it very seriously. It would be remiss of me not to listen to the athlete when they say: 'This is a bit of worry, I'm concerned about this.'"

The anxiety among the visitors sparked a series of health-care measures to be put in place by the Olympics committee and local government. Also the health-care advisories suggested using mosquito repellents and restricting sexual transmission by the use of condoms. Although the probability of worldwide Zika virus spread as a result of Rio Olympics 2016 was real, there are no reports suggesting a surge in the Zika virus cases in any of the participating countries (Figs. 3.5 and 3.6).

FIG. 3.5   Cartoon image portraying Olympic runners fleeing from a mosquito infested with Zika virus disease. *Acquired from: http://mediaspecialistsguide.blogspot.com/2016/08/the-rio-2016-resources- try-these-29.html.*

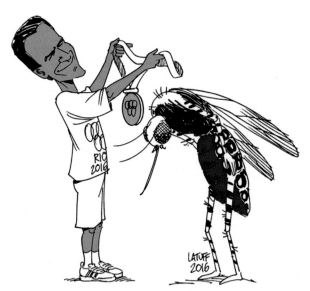

FIG. 3.6   Cartoon image of a mosquito being given the Olympic gold medal symbolizing its victory in spreading fear of Zika virus disease during the 2016 Rio Olympics. *Acquired from: http:// thehilltalk.com/2016/06/11/zika-virus-starts-to-scare-off-olympic- participants/.*

REFERENCES

1. History.com Staff. First victim of Memphis yellow-fever epidemic dies. Available from: http://www.history.com/this-day-in-history/first-victim-of-memphis-yellow-fever-epidemic-dies [accessed 9 January 2017].
2. Bloom KJ. *The Mississippi Valley's great yellow fever epidemic of 1878.* Baton Rouge: Louisiana State University Press; 1993. Rousey DC. Yellow fever and black policemen in memphis: a post-reconstruction anomaly. *J South Hist* 1985;**51**:357–74.
3. Wrenn LB. The impact of yellow fever on Memphis: a reappraisal. *West Tenn Hist Soc Pap* 1987;**41**:4–18.
4. Brink S. Yellow fever timeline: the history of a long misunderstood disease. Available from: http://www.npr.org/sections/goatsandsoda/2016/08/28/491471697/yellow-fever-timeline-the-history-of-a-long-misunderstood-disease [Accessed 9 January 2017].
5. Patterson KD. Yellow fever epidemics and mortality in the United States, 1693–1905. *Soc Sci Med* 1992;**34**:855–65.
6. Arnbeck B. A short history of Yellow fever in the US. Available from: http://bobarnebeck.com/history.html [Accessed 3 January 2017].
7. Gaunt MW, Sall AA, de Lamballerie X, et al. Phylogenetic relationships of flaviviruses correlate with their epidemiology, disease association and biogeography. *J Gen Virol* 2001;**82**:1867–76.
8. Calisher CH, Gould EA. Taxonomy of the virus family Flaviviridae. *Adv Virus Res* 2003;**59**:1–19.
9. Pletnev A, Gould E, Heinz FX, et al. Flaviviridae. In: King AMQ, Adams MJ, Carstens EB, Lefkowitz EJ, editors. *Virus taxonomy—ninth report of the International Committee on Taxonomy of Viruses.* Oxford: Elsevier; 2011. p. 1003–20.
10. Mukhopadhyay S, Kuhn RJ, Rossmann MG. A structural perspective of the flavivirus life cycle. *Nat Rev Microbiol* 2005;**3**:13–22.
11. Chu JJH, Ng ML. Trafficking mechanism of west Nile (Sarafend) virus structural proteins. *J Med Virol* 2002;**67**:127–36.
12. Kuhn RJ, Zhang W, Rossmann MG, et al. Structure of dengue virus: implications for flavivirus organization, maturation, and fusion. *Cell* 2002;**108**:717–25.
13. Mukhopadhyay S, Kim BS, Chipman PR, Rossmann MG, Kuhn RJ. Structure of West Nile virus. *Science* 2003;**302**:248.
14. Chambers TJ, Hahn CS, Galler R, Rice CM. Flavivirus genome organization, expression, and replication. *Annu Rev Microbiol* 1990;**44**:649–88.
15. Zhang Y, Zhang W, Ogata S, et al. Conformational changes of the Flavivirus E Glycoprotein. *Structure* 2004;**12**:1607–18.
16. Allison SL, Schalich J, Stiasny K, Mandl CW, Kunz C, Heinz FX. Oligomeric rearrangement of tick-borne encephalitis virus envelope proteins induced by an acidic pH. *J Virol* 1995;**69**:695–700.
17. Elshuber S, Allison SL, Heinz FX, Mandl CW. Cleavage of protein prM is necessary for infection of BHK-21 cells by tick-borne encephalitis virus. *J Gen Virol* 2003;**84**:183–91.
18. Allison SL, Stadler K, Mandl CW, Kunz C, Heinz FX. Synthesis and secretion of recombinant tick-borne encephalitis virus protein E in soluble and particulate form. *J Virol* 1995;**69**:5816–20.
19. Gebhard LG, Filomatori CV, Gamarnik AV. Functional RNA elements in the dengue virus genome. *Viruses* 2011;**3**(9):1739–56.
20. Brinton MA, Fernandez AV, Dispoto JH. The 3'-nucleotides of flavivirus genomic RNA form a conserved secondary structure. *Virology* 1986;**153**:113–21.

21. Holden KL, Harris E. Enhancement of dengue virus translation: role of the 3' untranslated region and the terminal 3' stem-loop domain. *Virology* 2004;**329**:119–33.

22. Markoff L. 5'- and 3'-noncoding regions in flavivirus RNA. *Adv Virus Res* 2003;**59**:177–228.

23. Modis Y, Ogata S, Clements D, Harrison SC. Structure of the dengue virus envelope protein after membrane fusion. *Nature* 2004;**427**:313–9.

24. Heinz FX, Allison SL. Structures and mechanisms in flavivirus fusion. *Adv Virus Res* 2000;**55**:231–69.

25. Ploubidou A, Way M. Viral transport and the cytoskeleton. *Curr Opin Cell Biol* 2001;**13**:97–105.

26. Villordo SM, Carballeda JM, Filomatori CV, Gamarnik AV. RNA structure duplications and flavivirus host adaptation. *Trends Microbiol* 2016;**24**:270–83.

27. Lai CJ, Pethel M, Jan LR, Kawano H, Cahour A, Falgout B. Processing of dengue type 4 and other flavivirus nonstructural proteins. *Arch Virol Suppl* 1994;**9**:359–68.

28. Ng ML, Howe J, Sreenivasan V, Mulders JJL. Flavivirus West Nile (Sarafend) egress at the plasma membrane. *Arch Virol* 1994;**137**:303–13.

29. Petersen LR, Marfin AA. Shifting epidemiology of Flaviviridae. *J Travel Med* 2005;**12**(Suppl 1):S3–11.

30. Hayes EB. Zika virus outside Africa. *Emerg Infect Dis* 2009;**15**:1347–50.

31. Althouse BM, Hanley KA, Diallo M, et al. Impact of climate and mosquito vector abundance on sylvatic arbovirus circulation dynamics in Senegal. *Am J Trop Med Hyg* 2015;**92**:88–97.

32. Haddow AJ, Williams MC, Woodall JP, Simpson DI, Goma LK. Twelve isolations Of Zika virus from Aedes (Stegomyia) Africanus (Theobald) taken in and above a Uganda forest. *Bull World Health Organ* 1964;**31**:57–69.

33. Marcondes CB, Ximenes MF. Zika virus in Brazil and the danger of infestation by Aedes (Stegomyia) mosquitoes. *Rev Soc Bras Med Trop* 2016;**49**:4–10.

34. Diagne CT, Diallo D, Faye O, et al. Potential of selected Senegalese Aedes spp. mosquitoes (Diptera: Culicidae) to transmit Zika virus. *BMC Infect Dis* 2015;**15**:492.

35. Musso D, Roche C, Robin E, Nhan T, Teissier A, Cao-Lormeau VM. Potential sexual transmission of Zika virus. *Emerg Infect Dis* 2015;**21**:359–61.

36. Mansuy JM, Dutertre M, Mengelle C, et al. Zika virus: high infectious viral load in semen, a new sexually transmitted pathogen? *Lancet Infect Dis* 2016;**16**:405.

37. Deckard DT, Chung WM, Brooks JT, et al. Male-to-male sexual transmission of Zika virus—Texas, January 2016. *MMWR Morb Mortal Wkly Rep* 2016;**65**:372–4.

38. Besnard M, Lastere S, Teissier A, Cao-Lormeau V, Musso D. Evidence of perinatal transmission of Zika virus, French Polynesia, December 2013 and February 2014. *Euro Surveill* 2014;**19**.

39. Duffy MR, Chen TH, Hancock WT, et al. Zika virus outbreak on Yap Island, Federated States of Micronesia. *N Engl J Med* 2009;**360**:2536–43.

40. Schuler-Faccini L, Ribeiro EM, Feitosa IM, et al. Possible association between Zika virus infection and microcephaly—Brazil, 2015. *MMWR Morb Mortal Wkly Rep* 2016;**65**:59–62.

41. Martines RB, Bhatnagar J, Keating MK, et al. Notes from the field: evidence of Zika virus infection in brain and placental tissues from two congenitally infected newborns and two fetal losses—Brazil, 2015. *MMWR Morb Mortal Wkly Rep* 2016;**65**:159–60.

42. Barthel A, Gourinat AC, Cazorla C, Joubert C, Dupont-Rouzeyrol M, Descloux E. Breast milk as a possible route of vertical transmission of dengue virus? *Clin Infect Dis* 2013;**57**:415–7.

43. Possible West Nile virus transmission to an infant through breast-feeding—Michigan, 2002. MMWR Morb Mortal Wkly Rep 2002;51:877–878.

44. Musso D, Nhan T, Robin E, et al. Potential for Zika virus transmission through blood transfusion demonstrated during an outbreak in French Polynesia, November 2013 to February 2014. *Euro Surveill* 2014;**19**:1–3.

45. Pealer LN, Marfin AA, Petersen LR, et al. Transmission of West Nile virus through blood transfusion in the United States in 2002. *N Engl J Med* 2003;**349**:1236–45.

46. Dauphin G, Zientara S. West Nile virus: recent trends in diagnosis and vaccine development. *Vaccine* 2007;**25**:5563–76.

47. Ishikawa T, Yamanaka A, Konishi E. A review of successful flavivirus vaccines and the problems with those flaviviruses for which vaccines are not yet available. *Vaccine* 2014;**32**:1326–37.

48. Attaran A. Off the podium: why public health concerns for global spread of Zika virus means that Rio de Janeiro's 2016 Olympic Games must not proceed [Internet]. Harvard Public Health Rev 2016.

49. Butler D. First Zika-linked birth defects detected in Colombia. *Nature* 2016;**531**:153.

Chapter 4

Global Healthcare Perspective

Chapter Outline

It would seem that many people around the world believed Zika virus to be a recently discovered virus. Zika virus infection reports never appeared on local news channels, families would not sit around the dinner table anxious about disease symptoms, and even healthcare professionals would view it as a rare disease that was of academic significance only. In fact, very little research was conducted on Zika virus since the initial discovery in 1947, and Zika virus infection was thought to be a disease that presented with mild, self-resolving symptoms, confined mostly to the African continent.[1] Of course, this perspective would change drastically in a span of a few years. What was practically unnoticed for more than half a century would cause such a devastating global impact, and it would force the global health community to mobilize vast amounts of time and resources. The disease would affect people from almost every continent, and would see the tireless efforts of health workers from all parts of the world, from local and national health organizations, to the World Health Organization (WHO) and United Nations Children's Fund.

This perceived instantaneous mobilization, however, was much slower than it should have been looking back on the sequence of events. As of March 9, 2017, 84 countries, territories, or subnational areas have been affected by the Zika virus infection. These areas have been identified using a new classification scheme developed by the WHO, the United States Centers for Disease Control and Prevention, and the European Centre for Disease Prevention and Control.[2] The classification also outlines recent countries and territories reporting microcephaly and other central nervous system malformations potentially associated with Zika virus infection. Systematic review of the literature up to May 30, 2016 has led the WHO and much of scientific community to the consensus

Zika Virus Disease. https://doi.org/10.1016/B978-0-12-812365-2.00005-6
63

that Zika virus infection can cause congenital brain abnormalities, including microcephaly, and that it is a trigger of Guillain-Barré syndrome.[3] So why has a virus with such devastating consequences been allowed to spread so quickly and expansively? Much of the explanation lies in the global community's early perception of this exotic virus and its strange and elusive history.

AN UNKNOWN VIRUS IS INTRODUCED TO THE WORLD

The ground is sticky, and muddy. It holds each step prisoner, just for a second. The air is humid. Mosquitos and moths are buzzing all around, and the forest is alive with sound. High above, the trees create a vast canopy to shield the world from the sun, and the overgrown foliage presents obstacles on every turn. The squelching underfoot is the sound of the swamp, but it is the screeching from above that is most noticeable. For it is up there, high up in the trees, where one can see flashes of brown among brilliant greens, that large platforms have been built to rest cages on. It is from one of these cages that the screeching emanates and on closer examination, the cage is home to a single rhesus monkey. But why is he there? And how did he come to be caged in his own native forest?

The answer to these questions lies with the members of the Yellow Fever Research Institute, a group established in 1936 by the International Division of the Rockefeller Foundation.[4] It is now April 1947, and scientists have come to this strange and remote forest in Uganda, Africa. The swamps of the Zika forest are supposed to reveal answers about Yellow Fever, and this rhesus monkey may provide those answers. He is sick with a fever and so he is isolated in the cage, and his blood is drawn for testing. Yet, what scientists find is completely unexpected. The serum of the monkey does test positive for a virus, but it is not the Yellow Fever virus. It is a virus that has never previously been isolated, a virus that would eventually be named for the forest it was discovered in. A year later, the same virus is again isolated in the same forest, this time from one of the abundant mosquitoes buzzing about, the *Aedes africanus*, and so the world is introduced to the Zika virus.[5]

Being introduced to a virus, however, does not mean that you are paying attention to it. The years following its discovery, the world would not see many occurrences of Zika virus infection in humans. That it not to say that knowledge of its capability to infect humans was unknown; that was most probably ascertained in the early 1950s.[1] In fact, the global health community would not put time and effort into researching various aspects of the virus, because in those rare human patients who were infected, it was a mild, self-resolving disease. It was verified in 1952 that Zika virus was in fact unique, and not identical to Yellow Fever or Dengue virus, but other than that, it almost seemed that the global health community believed Zika virus did not warrant further scientific inquiry.[1] Understandably, with such a small sample size, misconceptions can be formed, and it would be more than half a century before the global health community would get a chance to reexamine the Zika virus in a much larger sample size (population affected). Even then, the serious and life-threatening complications of Zika virus infection were not elucidated, and by the time the virus reached the shores of South America it would be too late to stop it.

THE FIRST OUTBREAK

Before 2007, there had been only 14 documented cases of human infection by Zika virus.[6] A group of physicians on Yap Island in the Federated States of Micronesia were about to change that. The residents of the island were experiencing an outbreak of an illness characterized by rash, conjunctivitis, and arthralgia. What made this particular outbreak strange was that, although serum from some of the sick patients had antibodies to Dengue virus, the presenting symptoms of this illness seemed different from Dengue fever. The physicians on the island decided to research the outbreak and discovered some surprising information. Nine out of 10 municipalities on Yap had patients with confirmed cases of Zika virus infection, and 73% of the population of Yap 3 years of age or older had been recently infected with Zika virus.[6]

This was the first real outbreak of Zika virus infection, and yet it was not met with any fear or trepidation from the global health community. The research conducted by the physicians indicated that the infection was a mild one without any serious complications.[6] To the world, Zika virus infection had now spread beyond Africa and Asia, but it was still a rare virus and, due to its relatively benign nature, was not cause for any concern. This could perfectly explain why Zika virus infection's devastating effect in South America, and consequently, other parts of the world, was not tempered by knowledge of the virus and its infectability. Or at least it would have been a perfect explanation, had there not been a larger and more serious outbreak of Zika virus infection beginning in 2013.

THE SECOND OUTBREAK

Every week the European Centre for Disease Prevention and Control publishes a Communicable Disease Threats Report. For the week ten report, March 2–8, 2014, a single paragraph was dedicated to an outbreak in some islands in the Pacific Ocean. The report outlined that an estimated 29 thousand cases of patients with Zika virus infection like symptoms had sought medical attention in the Island of French Polynesia since October 2013. And as of February 21, 2014, more than 8500 cases were suspected to actually be the Zika virus infection. Interestingly enough, 41 of those cases presented with Guillain-Barré syndrome. The report also noted the observation by French Polynesian health officials, of a concurrent significant increase in neurological syndromes since the start of the outbreak. Furthermore, the cause of this increase and its possible link to Zika virus were being investigated. The section of the report concluded by including updates of the other two islands experiencing an outbreak, another French territory, New Caledonia, and Easter Island.[7] The mention of the latter is of special significance, as it is a territory of Chile, and the focal point of one of the strongly believed theories of how Zika virus arrived in South America. Regardless, the information in this report actually represented another significant realization. The global health community was not unaware of the possible connection between Zika virus exposure and severe neurological complications. Yet, as we see in the early response to Zika virus infection in South America, the world seemed content to ignore this possibility, continuing to treat the Zika virus infection as a mild infectious disease.

One month after the publication of this specific Communicable Disease Threats Report, an article was published in *Eurosurveillance*, a European infectious disease journal. It details a case study of two pregnant patients in French Polynesia, who may have transmitted their Zika virus infection to their babies.[8] Yet another indication that, at least in Europe, the possibility of Zika virus effecting unborn fetuses should not have come as a surprise. Truly then, the reaction time to limit the spread of Zika virus infection and its complications was stunted not by a lack of exposure to Zika virus and its effects, but by preconceived notions that it was not a threat.

ZIKA VIRUS ARRIVES IN BRAZIL

On March 29, 2015 Brazil notified the WHO of an illness spreading across its northern states.[1] Between February and March 2015, nearly 7000 thousand cases were reported of patients presenting with mild skin rash, and at that point 425 blood samples had been taken.[1] Other than Dengue virus exposure, which was identified 13% of the 425 samples tested positive, there were no identifiable infectious agent.[1] Chikungunya, Measles, Rubella virus, Parvovirus B19, and Enterovirus were all excluded by serological tests. Although, at the time, no one would connect these cases to the unique virus from a forest in Uganda, these were in fact the earliest Zika virus infection cases in Brazil, preluding the rapid spread of the disease and its debilitating aftermath throughout the continent.

On May 7, 2015, the Brazil Reference Laboratory released information to the WHO finally and unequivocally confirming the presence of Zika virus infection in Brazil.[1] It had already been approximately 3 months since the arrival of the Zika virus infection on the shores of South America, or at least Brazil. Not much had been done after the first report, but now that Zika virus infection had been confirmed, the global health community finally turned its attention to Brazil. The World Health Organization, along with its regional office, the Pan American Health Organization, released the following statement in an epidemiological alert for the region:

> *The Pan American Health Organization (PAHO)/World Health Organization (WHO) recommends its Member States establish and maintain the capacity for Zika virus infection detection, clinical management and an effective public communication strategy to reduce the presence of the mosquito that transmits this disease, particularly in areas where the vector is present.*

Ref. 9.

The body of the alert briefly discussed Zika virus infections history and even included the neurological complications that occurred in French Polynesia. It also detailed its recommendations for member countries. There were seven areas of focus: surveillance, laboratory detection, case management, prevention and control measures, integrated vector management, personal prevention measures, and travelers. The alert was very comprehensive but in referring to case management, the alert signified that the need to differentiate Dengue virus from Zika virus disease was because Dengue, not Zika virus infection, had more severe clinical outcomes. It stated that: "Compared with Dengue, Zika virus infection has a more mild to moderate clinical picture, the onset of fever is more acute and shorter in duration; in addition, no shock or severe bleeding has been observed."[9]

The report is a snapshot of the global health perspective on Zika virus infection. The realization at this point was that Zika virus infection presents a problem because of its ability to spread expansively and quickly for two main reasons: the abundance of mosquitos and the abundance of travel into and out of the area. Brazil and its rainforests houses untold numbers of Aedes mosquitoes and the summer Olympics were fast approaching. No one thought about neurological complications or congenital malformations. Yet, even with all these precautions, the virus spread and revealed its true nature. The reason for this? The virus had already reached other countries of South America, and had already infected thousands in Brazil. Precautionary measures would not stop its march through South America.

ZIKA VIRUS INFECTION REVEALS ITSELF

In early July 2015, WHO was informed of 49 cases of Guillain-Barré syndrome, reported in Brazil, mostly from the state of Bahia.[1] All but two of those cases had a prior history of infection with Zika, Chikungunya, or Dengue virus infection; this was just the beginning. As the summer progressed, physicians observed more and more unexplainable neurological complications.

In the city of Recife, starting in August, physicians began seeing an unusual number of babies born with microcephaly, in some instances, seven babies in 1 day. Some of the mothers reported having a rash in early pregnancy. In October, Recife saw nearly 300 babies born with microcephaly. The Ministry of Health was informed, and after reviewing birth certificates they confirmed that there was in fact an increase in the number of births with microcephaly.[10] The ministry also opened a microcephaly registry for physicians and hospitals to report recent cases. Outside of Recife, more areas of Brazil were also reporting cases of microcephaly.[1]

By the end of November, Zika virus infection had spread to Suriname, El Salvador, Guatemala, Mexico, Paraguay, and Venezuela.[1] Outside of Brazil, there were as of that moment, no reports of microcephaly or neurological complications. Still, the virus was spreading like wildfire, and the WHO was trying to do everything to contain the spread. In its weekly epidemiological report on November 6, 2015, the WHO stated "Recent outbreaks of ZIKV [Zika virus] infection in different regions of the world underscore the potential for the virus to spread further in the Americas and beyond, wherever the vector is present."[11] The report encouraged countries to strengthen laboratory capacities to confirm cases of Zika virus infection, establish a surveillance system for the detection of neurologic complications, and strategize ways to engage communities to help reduce mosquito population.

Despite these efforts, as numbers of suspected cases of microcephaly continued to increase in Brazil, the Ministry was forced to declare a national public health emergency on November 11, 2015.[1] Five days later it reported the detection of Zika virus in the amniotic fluid samples of two pregnant women, whose fetuses were confirmed by ultrasound examinations to have microcephaly.[1] Thus, providing the first of many clues to the connection of microcephaly to Zika virus exposure. The same day the WHO and the Pan American Health Organization issued another alert asking countries to report increases in microcephaly or other neurological malformations. And still the Zika virus infection marched on.

As the months progressed, increasing number of countries were reporting Zika virus infections and their accompanying complications. In late November, Brazil reported three deaths, two adults and a newborn, related to the Zika virus infection.[1] By January 2016, both El Salvador and Brazil reported increasing numbers of cases of Guillain-Barré, and in El Salvador, two of those cases resulted in death.[1] In mid-January, the Hawaii Department of Health reported a baby with microcephaly, born to a mother who had resided in Brazil during early pregnancy.[1] As January closed out, Venezuela would join the ranks of countries with increased numbers of Guillain-Barré cases, bringing the total number to 252 associated both in time and place with Zika virus infection.[1] All the while Zika virus infection continued to spread, and by February it had encompassed almost all of South and Central America, including the Caribbean. February 2016 would herald another shocking revelation; the United States would report the first case of sexual transmission of Zika virus infection in Texas.[1]

Of course, by this point, the global health community was no longer a complacent onlooker to the crisis of Latin America. At the end of November 2015, the WHO and Pan American Health Organization would send a team of international experts to Brazil to assess the situation by reviewing epidemiological information, clinical data, laboratory capacity, and infrastructure of the country. The experts also provided input on case-control studies.[1] In January 2016, the American Center for Disease Control would collaborate with Brazilian health officials and release laboratory findings of four microcephaly cases that indicated the presence of Zika virus RNA (ribonucleic acid) through PCR (polymerase chain reaction).[1] This report would be one of the strongest piece of evidence linking Zika virus infection to microcephaly.

A PUBLIC HEALTH EMERGENCY OF INTERNATIONAL CONCERN

It had only been 9 months since Brazil's National Reference Laboratory confirmed the presence of Zika virus infection in the country and made the connection to outbreaks of strange illnesses in the north, but in such a short time Latin American discovered the very sinister nature of this disease. Many investigators were starting to suspect that Zika virus was the cause of an unusual increase in the number of cases of microcephaly, congenital malformations, and neurological complications in the region. As more and more evidence began to pile up in support of this association, the WHO decided to take action.

It is the evening of February 1, 2016 in Geneva, Switzerland, and a conference room is buzzing, ironically almost like mosquitoes, at the headquarters of the WHO. Crowded around the table are journalists, fingers poised over their laptop keys in anticipation, and many stand with pens hovering just above their pads. At the head of the conference table Dr. Margaret Chan, the Director General of the WHO, and Professor David Heymann, Chair of the Emergency Committee on Zika virus infection are sitting. As Dr. Chan prepares to speak, the room falls into an immediate silence. This is the press conference that will change the global heath community's approach to Zika virus infection.

During the February 1 conference, the WHO finally declared Zika virus infection a Public Health Emergency of International Concern, or PHEIC. Dr. Chan summarized the events of a teleconference meeting of the Zika Virus Emergency Committee that had occurred hours earlier. A committee was convened by Dr. Chan under International Health Regulations (2005), a form of international law. A panel of 18 experts and advisors had given their input on the threat of the Zika virus infection and specifically the suspected relationship of Zika virus infection to the rise in congenital malformations and neurological complications. The committee believed that there was a link, in both time and space, between infection by Zika virus and congenital malformations, although it could not be scientifically proven at that point. They advised that clusters of microcephaly in Latin America and neurological complications previously (in 2014 in French Polynesia) represented extraordinary events and that Zika virus infection does indeed pose a public health threat to other parts of the world. Furthermore, the committee recommended that the best approach would be a coordinated international response that would address certain measures. Dr. Chan outlined the most important measures, such as vector control, better methods of detection, and expedited development of vaccines.[1]

During the press conference the chairman of the Emergency Committee, Professor Heymann, also addressed the media. His approach was a little more cautious in the sense that he emphasized that the relationship between Zika virus infection and microcephaly and other neurological complications still could not be scientifically proven, but he did agree with the reason to declare a PHEIC as it is important to be preemptive. He explained that the committee's two main recommendations were to standardize surveillance of microcephaly and neurological complications in Zika virus-infected regions, and to encourage more research to establish a definitive scientific link.[1] Following the declaration of PHEIC, the global health community began to take a more preemptive approach to the Zika virus infection situation, and finally a slow but definitive response began to manifest.

THE STRATEGIC RESPONSE FRAMEWORK AND JOINT OPERATIONS PLAN

Two weeks after the press conference in Geneva, the WHO released a plan to guide the international response to Zika virus infection and its associated congenital malformations and neurological complications. The plan included development of guidance documents covering all aspects of response, translated into relevant languages. The global response to Zika virus infection would be coordinated from WHO headquarters. The organization would activate an incident management system in its headquarters and regional offices, which would allow for dedicated incident managers to draw on expertise and resources across the organization. The plan would include partnerships with other local and international organizations. This is how the Strategic Response Framework and Joint Operations Plan was born.[3]

The result of the PHEIC was a massive mobilization of international efforts, including the one above. The steps taken by the plan would at least make the world more aware of and prepared for Zika virus infection. These steps did not mean that the spread of Zika virus infection and its complications would stop. In fact, the United States would report two cases of Guillain-Barré syndrome associated with Zika virus infection and Panama would report one case in March 2016.[1] Still, the world was no longer complacent. From a global health perspective, Zika virus infection was not a mild infection from a forest in Africa anymore. It was a great threat that needed to be extinguished.

In June of 2016, the WHO launched a revised Zika virus infection strategic response plan that put even more focus on preventing and managing Zika virus infection's medical complications, researching the virus's effects, and creating stronger organizational partnerships. In one such partnership, the United Nations Children's Fund helped Brazil by supporting the Ministry of Health and Ministry of Education to promote communication and community mobilization through social media. In mid-March, the WHO convened a vector control advisory group to review available tools and to come up with new ways to control *Aedes* mosquito populations.

Because Brazil was the unfortunate focal point of Zika virus infection related attention from the global health community, another question arose. Could Brazil really host the 2016 Summer Olympic games? Yet again, here was an example of how the new outlook toward Zika virus infection allowed the implementation of an informed decision. The WHO addressed the issue in a press release:

Based on current assessment, cancelling or changing the location of the 2016 Olympics will not significantly alter the international spread of Zika virus.

Ref. 12.

Although this statement may not seem significant, it showed a much better understanding of the virus and the potential for infection. The world was now prepared, and people knew what precautionary measures to take, measures that were reemphasized in the rest of the press release. This included recommending that pregnant women refrain from traveling to areas with ongoing Zika virus infection.[12] At the beginning of the Zika virus infection epidemic, the WHO may have made the same decision, but only because such infection was not thought to be dangerous and so the necessary precautions would not have been taken, leading to devastating consequences. The global health view of Zika virus infection had changed, for the better.

AFTERMATH

Though many steps and measures have been taken since that fateful press conference in February 2016, not all have been outlined here. The Zika Virus Infection Emergency Committee met five times, and in November of 2016, on the basis of their recommendation, Dr. Chan would lift the PHEIC.[13] This did not mean the end of work on various aspects of Zika virus. Research, the development of vaccines, and vector control would all continue. The most recent count, as mentioned earlier, places 84 distinct territories as having pervious or ongoing Zika virus infections. The classification however was developed by the collaboration of three separate global health organizations; a feat that indicates a surprising fact about the Zika virus infection. What was once just a strange, rash causing, mild, self-limiting virus of the African continent, now had three of the largest health organizations monitoring its spread. The global health perspective on Zika virus infection has drastically changed, but that change will allow the world to combat this virus and its devastating medical complications.

REFERENCES

1. The history of Zika. World Health Organization. Accessed 8 February 2017, at http://www.who.int/emergencies/zika-virus/history/en/.
2. Situation report: Zika virus, microcephaly, Guillain-Barré syndrome. World Health Organization, 2017. Accessed January 2017, at http://www.who.int/emerg-encies/zika-virus/situation-report/10-march-2017/en/.
3. Zika virus fact sheet. World Health Organization. Accessed 10 March 2017, at http://www.who.int/media-centre/factsheets/zika/en/.

4. History of Health Research at UVRI. Uganda Virus Research Institute. Accessed 1 February 2017 http://www.uvri.go.ug/in-dex.php/about-uvri/history.

5. Wright C. Even in the place where Zika virus was first discovered, its true origin is a mystery. Quartz. https://qz.com/635161/even-in-the-place-where-zika-virus-was-first-discovered-its-true-origin-is-a-mystery/. Published May 5, 2016. Accessed 1 February 2017.

6. Duffy MR, Hanckock WT. Zika virus outbreak on Yap Island, Federated States of Micronesia. *N Engl J Med* 2009;**360**(24):2537–43.

7. Communicable disease threat report. ECDC, 2014. Accessed 7 March 7, at http://ecdc.europa.eu/en/publications/Publications/communicable-disease-threats-report-08-mar-2014.pdf.

8. Roth A, Mercier C. Concurrent outbreaks of dengue, chikungunya and Zika virus infections—an unprecedented epidemic wave of mosquito-borne viruses in the Pacific 2012–2014. *Eurosurveil* 2014;**19**(41). http://www.eurosurveillance.org/ViewArt-icle.aspx?ArticleId=20929.

9. Epidemiological alert: Zika virus infection. World Health Organization, 2015. Accessed 7 April 2017, at http://www2.paho.org/hq/index.php?option=com_doc-man&task=doc_view&Itemid=270&gid=30075=en%20%28accessed%2002%20feb%202016%29.

10. Sifferlin A. How Brazil uncovered the possible connection between Zika and Microcephaly. Time. http://time.com/4202262/zika-brazil-doctors-recife-investigation-outbreak/. Published February 1, 2016. Accessed 3 February 2017.

11. Epidemiological alert: Zika virus outbreaks in the Americas. World Health Organization, 2015. Accessed 7 April 2017, at http://www.who.int/wer/2015/wer9045.pdf?ua=1.

12. WHO public health advice regarding the Olympics and Zika virus [Press Release]. World Health Organization, 2016. Accessed 4 February 2017, at http://www.who.int/media-centre/news/releases/2016/zika-health-advice-olympics/en/.

13. Fifth meeting of the Emergency Committee under the International Health Regulations (2005) regarding microcephaly, other neurological disorders and Zika virus. World Health Organization, 2016. Accessed 4 February 2017, at http://www.who.int/media-centre/news/statements/2016/zika-fifth-ec/en/.

Chapter 5

United States Health Care Perspective

There was no a massive death toll. People did not attend funerals only to return home, get sick, and begin a downward spiral towards death. It was nothing like the monstrous Ebola virus disease epidemic. Yet, there was palpable panic in the United States as the scourge of the Zika virus infection crept slowly, but surely, to its shores. Zika virus infection had not been a problem—or at least not a recognized problem—in the United States and the public did not anticipate any danger in the early throes of the South American epidemic, even if, in retrospect, it would have been prudent to be better prepared. The biggest threat, for a lot of Americans, therefore, only came from traveling to countries where the virus was ravaging the population. This meant fewer people from the United States would travel to Brazil for the Rio Olympic Games in July 2016.

According to one estimate, the expected attendance rate from the United States dropped by 50% from 200,000 to 100,000, in light of the Zika virus infection epidemic.[1] Americans, who would have usually made a livelihood out of providing security and other services to fellow American visitors in Brazil, faced major shortfalls. According to an article published in June 2016, Airbnb declared that the "safety of the Airbnb community is the single most important thing we work on every day."[1] In fact, Airbnb emphasized the importance of following the guidelines of the World Health Organization and the United States Centers for Disease Control and Prevention. Coca-Cola provided a similar statement to its employees, excusing them from attending the Rio Games.[1] Security firms were devastated in the strangest manner.

A security firm run by a retired Federal Bureau of Investigation agent had established an understanding with Brazilian security teams to acquire "fleets of armored cars and buses to ferry visitors," around Brazil during their stay.[1] However, no amount of armor was sufficient to comfort and lure potential travelers, and these security firms worried that their cars would "sit unused."[1] The security Americans were looking for could not be provided by traditional security firms. The threat was from a mosquito and the virus. This fear persisted despite the announcement of Brazil's Health Minister that people had close to "zero risk" of acquiring Zika virus infection in Rio because the games were taking place in July, a winter month in South America when the mosquito population would be low. Despite the fact that political unrest is a source of great

Zika Virus Disease. https://doi.org/10.1016/B978-0-12-812365-2.00006-8

danger in many South American countries, according to a report in USA Today, most Americans still continued to cite the Zika virus infection as their primary concern when considering travel to the Rio Games.[1]

The Centers for Disease Control and Prevention had already activated its Emergency Operations Center, which had been operationalized 10 years earlier as an emergency response mechanism in response to the anthrax and World Trade Center attacks.[2] Through the Emergency Operations Center, the Centers for Disease Control and Prevention, in January of 2016, had begun developing diagnostic tests for Zika virus infection, studying any links between Zika virus infection and microcephaly and Guillain-Barre syndrome, monitoring and reporting cases of Zika virus infection , providing guidance to travelers, and carrying out surveillance in the United States and its territories.[2] In fact, on February 8, 2016, the Centers for Disease Control and Prevention raised its Emergency Operations Center activation to the highest level.[2] If the Zika virus infection was not as dreadful in its death toll or as glum in outlook as the Ebola virus disease, and the Centers for Disease Control and Prevention was already responding, then why were so many Americans concerned about traveling to the Rio Games? A new mother in the United States, who had been living in Brazil prior to giving birth, might be the representation of the underlying concerns.

About 6 months earlier, on January 16, 2016, a *New York Times* article announced to the world that the United States had its first case of brain damage in a newborn tied to Zika virus infection.[3] That Friday, in an Oahu hospital in Hawaii, a woman had given birth, as many mothers before her had done since time immemorial. However, the child this mother gave birth to suffered from microcephaly, "an unusually small head and brain."[3] The Centers for Disease Control and Prevention confirmed the presence of Zika virus infection in this newborn. The question remained: how did this happen in the United States? The Hawaii State Department of Health provided its understanding of the situation.

The mother had lived in Brazil in May 2015, where she was presumably infected by a mosquito early in her pregnancy.[3] The virus was postulated to have reached the embryo and damaged the brain.[3] Along with expressing her sadness at this outcome, Hawaii's state epidemiologist took the opportunity to emphasize, "the importance of Centers for Disease Control and Prevention travel recommendations" released that day.[3] In addition to reiterating travel guidelines, Hawaii benefited from an existent "Fight the Bite" campaign. Intended to address Dengue virus infection, this campaign encouraged residents to try to avoid getting bitten by using mosquito repellant, wearing long sleeves, and by getting rid of all standing water on their properties.[3] Since Dengue virus is spread by the same vector, this campaign was seen as useful to prevent spread of Zika virus infection as well. At the time of this writing, Hawaii had 16 symptomatic cases of Zika virus infection, which was much lower than the thousands of cases in Florida and Puerto Rico and the hundreds of cases in Texas.[4]

Between the reports of microcephaly in a baby whose mother had traveled to Brazil, and seeing pictures of the children with microcephaly, there was enough of a deterrent for most people to refrain from travelling to the Rio Games. Presumably due to stringently applied federal patient protection laws provided by the Health Insurance Portability and Accountability Act (HIPPA), very few, if any, images of babies born in the US with microcephaly suspected to be caused by Zika virus infection have been published in newspapers. Therefore, the babies afflicted in South America would have to serve as "poster babies" to demonstrate the urgency and seriousness of Zika virus infection. Supplementing these images and the report of the first United States baby affected by Zika virus infection was also the fact that on February 1, 2016, the World Health Organization had declared Zika virus infection a Public Health Emergency of International Concern.[5] These factors contributed to the fear of contracting Zika virus infection while traveling to Rio or to any South American country in the midst of the outbreak.

If you ask an infectious disease epidemiologist what an outbreak is or how many cases are required to label something an outbreak, he or she would tell you that even one case can be called an outbreak if it is an unusual occurrence for the time and/or place. If you live in New York State in the United States and never traveled out of your city but one day fall ill, end up in the hospital, and get diagnosed with malaria, an astute infectious disease epidemiologist would consider you an outbreak case. That is because malaria does not exist in the United States. The only way a person can acquire malaria is if he or she travels to a country where it is common and returns to the United States with the disease. Just as Zika virus infection was declared a Public Health Emergency of International Concern in February of 2016, a disturbing development came out of Dallas, Texas.

Just one day after the World Health Organization had declared a Public Health Emergency of International Concern, Reuters reported that a person in Dallas, Texas, had developed Zika virus infection without traveling to South America.[6] This person had been in sexual contact with someone who had traveled to Venezuela. Centers for Disease Control and Prevention director, Dr. Tom Frieden, labeled this as the first United States-based case of Zika virus infection in someone who had not traveled abroad.[6] This was an alarming finding and was considered a newer route of transmission. It was just one case, but given the unusual nature of transmission it qualified for outbreak investigation and surveillance. Through this revelation, the Centers for Disease Control and Prevention was spurred to publish guidelines on safe sex practices. Based on available data, the guidelines warned that the Zika virus infection could remain in the semen of a male for much longer after it was no longer detectable in the blood. However, no diagnostic testing was recommended for men at the time due to the unclear understanding of the duration of presence of the virus in the male genitourinary tract.[7] An especially meticulous Cable News Network (CNN) journalist, however, dispelled the notion that this was the first case of sexual transmission of Zika virus infection in the United States.[8]

About 8 years earlier, a potential sexual transmission of Zika virus infection had been studied, identified, and a report had been published in the medical literature.[8]

However, the scientific community wrote off the finding as "inconclusive," a "one-off case," or "unsubstantiated."[8] It was only in September 2016, in the *Bulletin of the World Health Organization*, that the 2008 case was listed as the first case of sexual transmission of Zika virus infection in the United States.[5] In fact, this case was also the first case of Zika virus infection ever identified in the United States. The patient who transmitted the Zika virus infection and the person who acquired it, both were well versed with disease investigation and so they documented and studied their own cases with scientific rigor. The affected couple even published a scientific paper on their findings. This case would bring to light the strange nature of scientific research that both propels us forward and holds us back.

The story began in the summer of 2008.[8] The main characters are Brian Foy, a professor at Colorado State University; his graduate student, Kevin Kobylinski; and Foy's wife, Joy Chilson Foy. The setting was first Senegal and then northern Colorado. Professor Foy and Kobylinski were studying ways to stop the spread of Malaria in Senegal.[8] This led them on long treks through villages known to be home to mosquitoes. Both the professor and his student were prime dining spots for hungry female *Aedes aegypti* mosquitoes. As part of their research they used vacuum aspirators to collect mosquitoes to study them.[8] During this process, they were probably inadvertently bitten. However, repelling mosquitoes were not their primary concern. After all, they had already taken antimalarial prophylactic medications and they had received all the necessary vaccines.[8] Professor Foy and Kobylinski returned to the United States. Professor Foy felt perfectly fine and was overjoyed to be back home in northern Colorado with his children and his wife, Joy.[8]

About 10 days later, however, Professor Foy began to feel ill with joint pain and swelling, headache, light sensitivity, and a rash on his back and chest. His initial instinct was to suspect a flavivirus, the family of viruses that can be transmitted by *Aedes aegypti* mosquitos that are prevalent in Senegal.[8] Despite the exhaustion he felt, he decided to try to figure out what was causing his illness. He began documenting the symptoms he was having and called his student to see if he was also suffering from similar complaints. At about the same time, Kobylinski had also come down with a rash and body aches.[8] The next crucial step in Professor Foy's mind was to have his blood and his student's blood drawn while they were actively experiencing symptoms so that the virus could be identified. It was at this time, that Joy, Professor Foy's wife, also fell ill.[8] Professor Foy was intrigued by this occurrence. He was looking for a mosquito-borne illness and his wife, who had never travelled to an area labeled endemic for such diseases was developing the same, albeit perceivably more severe, symptoms: headache, joint pains, light sensitivity, and rash.[8] Additionally, northern Colorado was not home to the *Aedes aegypti* mosquito.

Joy, being a registered oncology nurse, also began tracking her symptoms.[8] Professor Foy observed new symptoms of his own. He began having prostate pain and blood in his semen; this clued him to a sexual link to the transmission of the virus.[8] Based on the couple's sexual history, Professor Foy predicted that the transmission occurred when he had first returned to the United States and was not even symptomatic. Professor Foy had sent blood samples from all three of them to the Centers for Disease Control and Prevention's Division of Vector-Borne Diseases in Fort Collins, Colorado.[8] Professor Foy's blood and Kobylinski's blood showed cross-reactivity to Dengue and Yellow Fever virus, which meant that they had been exposed to or vaccinated against these diseases. But Joy's blood did not reveal anything. Zika virus exposure had not been tested for because it was too unusual and not well known.[8] With no new information, Professor Foy stored the blood samples in his laboratory and, having recovered from his illness, continued his normal life.[8] Unbeknownst to him, this was not the end of his story.

Not allowing the painful experience of the previous year to get in the way of science research, Kobylinski returned to Senegal to carry forward the malaria research in other villages. This was where Kobylinski ended up having lunch with Andrew Haddow, a researcher from the United States Army Medical Research Institute of Infectious Diseases.[8] The United States Army research facilities have vast experience and resources available at their disposal, and Haddow was working with fellow researchers to study rare viruses that affected people in the region.[8] Learning this, Kobylinski mentioned the experience he, Professor Foy, and Joy had had. Upon hearing the symptoms and the results, Haddow told Kobylinski that it could potentially have Zika virus infection.[8] This was no off-the-cuff hypothesis. Haddow was the grandson of Alexander Haddow, the man who had discovered the Zika virus infection in the Zika forest in Uganda.[8]

Kobylinski shared this hypothesis with Professor Foy, and Foy was excited with the prospect of further studying the disease that had, a year earlier, mysteriously engulfed him and his wife. He sent the frozen blood samples to Haddow, who had them processed at the University of Texas Medical Branch in Galveston, Texas.[8] All three samples returned positive for Zika virus infection, and a second test confirmed the results. These tests, though, relied on antibodies and the actual virus was not isolated, which is what led some in the scientific community to disregard these findings.[8] This was a case in which meticulous concepts of infectious disease investigation were followed and where the scientific community reacted with an expected dose of caution. This caution, bordering on unfounded skepticism, may have ultimately led to the delayed recognition of the sexual transmission of Zika virus infection until the case in Dallas, Texas, 8 years later. Professor Foy, upon publishing his findings, had submitted a grant to the National Institutes of Health (NIH) to get funding to study this phenomenon further, but the grant was not accepted.[8] In the end, though, the trio seemed to be isolated cases of Zika virus infection as their occurrence was far removed from the first case, during the current epidemic, that developed in a United States territory or state.

On December 31, 2015, the United States experienced its first Zika virus infection outbreak of the current epidemic,[5] although not on the mainland.

This was before the World Health Organization declared a Public Health Emergency of International Concern, before the Centers for Disease Control and Prevention had activated its Emergency Operations Center, and before the Centers for Disease Control and Prevention had developed a case definition. The case was confirmed by polymerase chain reaction (PCR)-testing and it was termed a "locally acquired" infection.[5] This meant that the person, who was infected, had not travel to another country that was experiencing a known outbreak, but rather was bit by a local mosquito. The territory was Puerto Rico. This is not unexpected because the *Aedes aegypti* mosquito can live in Puerto Rico, where the environment is perfect for its survival. The Centers for Disease Control and Prevention immediately sent out a press release that reported the finding and emphasized travel precautions.[9] With no vaccine available and no known cure, the best approach that anyone could offer was prevention: wearing long sleeve shirts, long pants, using insect repellent, putting up screens on windows and doors, and getting rid of standing water—a breeding ground for mosquitoes.[9]

Almost a year after the first case, Puerto Rico would see more than 34,000 confirmed cases of Zika virus infection and more than 2600 pregnant women had been tested positive for Zika virus infection.[10] At this point in the timeline of the epidemic, a link between Zika virus infection and microcephaly was not yet regarded as a confirmed association. In the rest of the United States, like in Hawaii, the cases of Zika virus infection were travel related.

Finally, by February 2016, enough cases of Zika virus infection had occurred in the United States and enough babies had been born with microcephaly that then President, Barack Obama, requested $1.8 billion in emergency funding to respond to and combat Zika virus infection.[11] Under this request, the Department of Health and Human Services would receive $1.48 billion of the total and $828 million was to be given to the Centers for Disease and Prevention for mosquito control programs, improving Zika virus infection diagnostics, and public health outreach—$200 million would go toward vaccine development and $250 million would go toward health services in Puerto Rico.[11] Unfortunately, this was only a request and it ended up getting delayed in Congress.

During this time, the cases of Zika virus infection continued to rise. By March 9, 2016 the United States had experienced 193 travel-associated Zika virus infection disease cases.[4] It was also only after a Brief Report published in the *New England Journal of Medicine*, in March 2016, that the scientific community accepted the association between Zika virus infection and brain damage in the fetus.[12] The Centers for Disease Control and Prevention prepared for more cases and was concerned that, similar to Puerto Rico, there could be an infection outbreak in the mainland United States that was not travel associated. On April 1, 2016, the Centers for Disease Control and Prevention held a meeting at its headquarters to better coordinate its response to the Zika virus infection.[13] More than 300 local, state, and federal authorities, and experts gathered for the meeting in the hopes of preparing for potential "clusters of mosquito-transmitted" Zika virus infections in the "US mainland."[13] During this meeting, experts and officials shared scientific knowledge, and received training on how to prevent and treat Zika virus infections, and how to communicate with the public, especially pregnant women, about the health effects of Zika virus infection.[13] The concern of locally acquired Zika virus infection stemmed from the fact that the mosquito that carries the Zika virus infection is also endemic in the southern United States, specifically Texas and Florida. The scarier fact was that the *Aedes albopictus* mosquito, which can also carry Zika virus infection, is present in more northern states as well. As these essential gatherings took place, scientists continued to put out warnings and the slow-moving machinations of Congress were finally propelled into some action.

After a 2-month struggle, Congress finally agreed to allocate $589 million to fight against Zika virus infection.[14] This was far short of the $1.8 billion requested by the White House, and of the $589 million dedicated, $510 million were being transferred from the $2.7 billion that had been reserved for the battle against Ebola virus disease. The remaining $79 million were being siphoned from funds designated for "emergency medical supplies during epidemics and national vaccine stockpiles."[14] This money would be used for vaccine development, diagnostics, and mosquito control. Meanwhile in the United Kingdom, a company called Oxitec began working to create genetically modified male mosquitoes that would mate with the female mosquitoes and lead to male-only offspring.[15] This would help control the spread of Zika virus infection, as it is the female mosquito that bites people and, thus, transmits the virus. With a decreasing female mosquito population, Zika virus infection could potentially be controlled. However, such a solution did not seem easily viable in the United States because of public concern and skepticism.[15] The Food and Drug Administration (FDA), Centers for Disease Control and Prevention, and Environmental Protection Agency (EPA) had already approved a proposal of Oxitec to test the model in the US in 2017.[15] People living in the United States, even in the states most affected by Zika virus infection, were more concerned with the unintended consequences of the genetically modified mosquitoes and their unknown effects on the ecosystem.[15]

Despite progress at one end, the Zika virus infection baffled scientists and healthcare providers. Aside from the devastating effects on fetuses, and the potential to cause Guillian-Barre syndrome in adults, Zika virus infection was not known to have a fatal effect. That was until April of 2016, when the first death from Zika virus infection in Puerto Rico was reported.[16] The patient, a man in his 70s, had acquired Zika virus infection and was hospitalized. His symptoms subsided, but that is when he began exhibiting bleeding, which is common in some other mosquito-borne viruses such as Dengue virus. This man was diagnosed and died with a rare complication of Zika virus infection—immune thrombocytopenic purpura (ITP), when the body attacks its own blood cells called platelets.[16] One unique characteristic in this patient's history was that he had had Dengue virus infection in the past. Scientists were left wondering whether a

Zika virus infection following a Dengue virus infection can be more severe.[16] In July 2016, there was another report of a "first United States Zika virus infection - related death." This time the victim was a resident of Salt Lake City, Utah. The 73-year-old had traveled to Mexico where he got infected by Zika virus infection and returned to the United States where he was being cared for at the University of Utah Health Care.[17] The medical team and the Centers for Disease Control and Prevention both agreed that the cause of the man's death was Zika virus infection. This man also had had a Dengue virus infection in the past.[17] In the same month that year, the fears of authorities would turn to reality.

On Friday, July 29, 2016, health officials in Florida announced that four individuals in the counties of Miami-Dade and Broward had been infected with Zika virus infection through transmission from local mosquitoes.[18] As 80% of people infected with Zika virus infection will show no signs of infection, Florida Health Department officials were going door-to-door, in the zip code of Miami affected by Zika virus infection, to collect urine samples from residents to determine how many people might be infected.[18] At this time, blood collection laboratories also began testing donated blood samples to ensure a clean supply.[18] Governor Rick Scott had already allocated $26.2 million in State funds to combat Zika virus infection .[19] At this juncture, he further emphasized the need for people to take precautions, such as using mosquito repellents, disposing of standing water, and getting tested for Zika virus infection. He further ordered pest control companies to increasing spraying and abatement efforts. Governor Scott also encouraged obstetricians/gynecologists to distribute Zika virus infection prevention kits to pregnant women.[19] Each of these kits consisted of a bed net, mosquito spray, standing water treatment tablets, permethrin spray for clothing, and condoms.[20]

The Centers for Disease Control and Prevention distributed a press release on August 1, 2016 in which it, along with the state of Florida, issued travel guidance, which recommended that pregnant women should not travel to the affected area in Florida.[21] However, this travel "ban," as some referred to it, was lifted in September after there were no new recorded cases of Zika virus infection transmission in the area for 45 days.[22] Still, measures to prevent the spread of Zika virus infection continued. At the request of Florida, the Centers for Disease Control and Prevention sent an Emergency Response Team with experts in Zika virus infection, "pregnancy and birth defects, vector control, laboratory science, and risk communications to assist in the response."[21] Despite such efforts, unfortunately, the spread of Zika virus infection continued. Close to 43 people had contracted Zika virus infection by the end of August.[23] Governor Scott, at this time, traveled to the south of Florida and held a "roundtable discussion" at which it was decided that the Miami Beach Botanical Garden would be closed for 2–3 days "in order to give the city and the county an opportunity to properly inspect and treat for mosquitoes."[23] By the end of 2016, Florida had had 218 symptomatic cases of locally acquired Zika virus infection.[24] Modelers and scientists had been correctly warning that the local spread of Zika virus infection was capable of reaching the shores of mainland United States, and this was just the beginning.

On Monday, November 28, 2016, Texas became the second state to have a case of locally acquired Zika virus infection. The first case of locally acquired Zika virus infection in Texas was found in Brownsville near the Mexican border. By the end of 2016, Texas had reported 6 cases of locally transmitted Zika virus infection. In the face of this growing epidemic, the Centers for Disease Control and Prevention and State health officials continued to promote prevention strategies. By this time, there had been 5102 cases of Zika virus infection in the mainland United States.[24] The response of people living in the United States was not rapid enough.

The Associated Press-NORC Center for Public Affairs Research conducted a poll in March 2016 on the awareness of people on Zika virus infection. According to the poll, 4 in 10 Americans had heard a little or nothing at all about the Zika virus infection.[25] Another study on the twitter reactions and responses of Americans to Zika virus infection might shed some light on the dearth of awareness. This study found that Twitter users preferred to look at user-generated links rather than governmental or official twitter links from Centers for Disease Control and Prevention.[26] Additionally, users wanted to know the impact of Zika virus infection on their health. Considering that Zika virus infection is asymptomatic in most, people were not inclined to "re-tweet" information on Zika virus infection.[26] In the end, the rigorous campaigns of the Centers for Disease Control and Prevention and state health departments brought about greater knowledge about Zika virus infection and its modes of transmission. A look at the number of reported symptomatic cases in the first half of 2017 reveals that there has been a decline in the number of Zika virus infection cases compared to the second half of 2016—119 versus 1502.[24] None of the 119 cases were the result of local transmission.

Since the Zika virus infection epidemic was not as threatening as the Ebola virus disease or the deadly polio virus epidemic of the 1920s and since there was some preparedness on part of the Centers for Disease Control and Prevention and State Health Departments, the Zika virus infection epidemic in the United States has remained contained. Time will tell if the public health efforts were truly successful in stamping out the epidemic and if the United States will be prepared to deal with other emerging infectious diseases.

REFERENCES

1. Gomez A. USA Today. Zika is among reasons many Americans skipping Rio Olympics, 2016; https://www.usatoday.com/story/sports/olympics/2016/06/22/americans-rio-summer-games-zika-virus/86143768/. Accessed December 2016.

2. CDC. Office of Public Health Preparedness and Response. Emergency Operations Centers: CDC Emergency Operations Center (EOC), 2016; https://www.cdc.gov/phpr/eoc.htm. Accessed December 2016.

3. Jr. DGM. The New York Times. Hawaii baby with brain damage is first U.S. case tied to Zika virus, 2016; https://www.nytimes.com/2016/01/17/health/hawaii-reports-baby-born-with-brain-damage-linked-to-zika-virus.html. Accessed 18 December 2016.

4. CDC. Zika virus. Case Counts in the US, 2017; https://www.cdc.gov/zika/geo/united-states. html. Accessed 3 May 2017.

5. Kindhauser MK, Allen T, Frank V, Santhana RS, Dye C. Zika: the origin and spread of a mosquito-borne virus. *Bull World Health Organ* 2016;**94**(9):675C–86C.

6. Herskovitz J. Reuters. First U.S. Zika virus transmission reported, attributed to sex, 2016; http://www.reuters.com/article/us-health-zika-idUSKCN0VB145. Accessed 20 December 2016.

7. Oster AM, Brooks JT, Stryker JE, et al. Interim guidelines for prevention of sexual transmission of Zika virus—United States, 2016. *MMWR Morb Mortal Wkly Rep* 2016;**65**:120–1. https://doi.org/10.15585/mmwr.mm6505e1.

8. LaMotte S. CNN. First known sexual transmission of Zika virus in U.S. was eight years ago, 2016; http://www.cnn.com/2016/02/17/health/first-zika-virus-sexual-transmission/. Accessed 8 January 2017.

9. CDC Newsroom. First case of Zika virus reported in Puerto Rico, 2015; https://www.cdc.gov/media/releases/2015/s1231-zika.html. Accessed 30 December 2016.

10. Beaubien J. NPR. Zika Pregnancies And Big Questions In Puerto Rico, 2016; http://www.npr.org/sections/health-shots/2016/11/29/503592597/zika-pregnancies-and-big-questions-in-puerto-rico. Accessed 14 February 2017.

11. Mole B. ARS Technica. Obama says no need to panic over Zika, requests $1.8 billion, 2016; https://arstechnica.com/science/2016/02/obama-says-no-need-to-panic-over-zika-requests-1-8-billion/. Accessed 5 March 2017.

12. Driggers RW, Ho C-Y, Korhonen EM, et al. Zika virus infection with prolonged maternal Viremia and fetal brain abnormalities. *N Engl J Med* 2016;**374**(22):2142–51.

13. Mole B. ARS Technica. CDC braces for Zika's US invasion as scientists watch virus melt fetal brain, 2016; https://arstechnica.com/science/2016/04/cdc-braces-for-zikas-us-invasion-as-scientists-watch-virus-melt-fetal-brain/. Accessed 15 January 2017.

14. Mole B. ARS Technica. After standoff with Congress, White House robs Ebola fund to pay for Zika, 2016; https://arstechnica.com/science/2016/04/after-stand-off-with-congress-white-house-robs-ebola-fund-to-pay-for-zika/. Accessed December 2016.

15. Bukszpan D. CNBC. The race is on to stop a Zika virus epidemic in the US, 2017; http://www.cnbc.com/2017/04/11/the-race-is-on-to-stop-a-zika-virus-epidemic-in-the-us.html. Accessed 5 April 2017.

16. Steenhuysen J. Reuters Health News. Puerto Rico Zika cases now include 65 pregnant women, one death: CDC, 2016; http://www.reuters.com/article/us-health-zika-puertorico-idUSKCN0XQ24R. Accessed 21 November 2016.

17. Branswell H. STAT. First Zika death in the US was indeed caused by the virus, officials say, 2016; https://www.statnews.com/2016/09/28/zika-death/. Accessed 13 April 2017.

18. Goldschmidt D. CNN. Florida health officials confirm local Zika transmission, 2016; http://www.cnn.com/2016/07/29/health/florida-health-officials-confirm-local-zika-transmission/. Accessed 15 April 2017.

19. Gov. Scott: With likely mosquito-borne Zika cases, state will use full resources to protect Floridians. 2016; http://www.flgov.com/2016/07/29/gov-scott-with-likely-mosquito-borne-zika-cases-state-will-use-full-resources-to-protect-floridians/. Accessed 12 January 2017.

20. CDC's Response to Zika. Pregnant and living in an area with Zika? Zika Prevention Kit for Pregnant Women, 2016; https://www.cdc.gov/zika/pdfs/zika-prevention-kit-english.pdf. Accessed 20 April 2017.

21. CDC. CDC Newsroom. CDC issues travel guidance related to Miami neighborhood with active Zika spread, 2016; https://www.cdc.gov/media/releases/2016/p0801-zika-travel-guidance.html. Accessed 25 December 2016.

22.   Gomez A. USA Today. CDC lifts travel ban as Miami neighborhood declared Zika free, 2016; https://www.usatoday.com/story/news/health/2016/09/19/miami-wynwood-declared-zika-virus-free/90682476/. Accessed 5 March 2017.

23.   Local ABC 10 News. Zika Virus. Governor back in South Florida as additional Zika virus transmission reported in Miami Beach: Miami Beach Botanical Garden temporarily closes due to Zika concerns, 2016; https://www.local10.com/health/zika-virus/governor-back-in-south-florida-as-additional-zika-virus-transmission-reported-in-miami-beach. Accessed 15 March 2017.

24.   CDC. Zika Virus 2016 Case Counts in the US, 2016; https://www.cdc.gov/zika/reporting/2016-case-counts.html. Accessed 5 May 2017.

25.   AP. The Zika virus: Americans' awareness and opinions of the U.S. response, 2016; http://apnorc.org/projects/Pages/HTML%20Reports/the-zika-virus-americans-awareness-and-opinions-of-the-us-response.aspx. Accessed 11 March 2017.

26.   Fu KW, Liang H, Saroha N, Tse ZT, Ip P, Fung IC. How people react to Zika virus outbreaks on Twitter? a computational content analysis. *Am J Infect Control* 2016;**44**(12):1700–2.

Chapter 6

Comparing the Zika Virus Disease Pandemic to Other Disease Pandemics

Chapter Outline

EPIDEMIC VERSUS PANDEMIC TERMINOLOGY

Epidemic and pandemic are terms primarily distinguished in terms of spread of contagious, infectious, or viral illness. An epidemic is limited to one specific region while a pandemic has a worldwide spread.[1] Zika virus disease is an epidemic which was reported in Brazil in Feb., 2016. Kindhauser et al.[2] reported temporal and geographical distribution of Zika virus disease from 1947 to Feb. 2016. During the aforementioned time period, about 74 countries and territories reported cases of human Zika virus infection.

Zika virus was first discovered in 1947 in Uganda, isolated from mosquitoes in 1948, first human infection reported in 1952, initial spread from Asia to Pacific Island in 2007, first known instance of sexual transmission in 2008, Guillain-Barré syndrome and microcephaly linked to Zika virus infection were reported in 2014 and 2015, respectively, and first infected persons reported in Americas in 2015.[2] Locally transmitted infections from Brazil were reported to the World Health Organization (WHO) in May 2015, followed by association of Guillain-Barré syndrome with Zika virus infection in July 2015. In October 2015 microcephaly in babies whose mothers had Zika virus infection during the pregnancy—but there was not causal link between these neurological complications and Zika virus infection

Zika Virus Disease. https://doi.org/10.1016/B978-0-12-812365-2.00007-X
85

established[2a]. In Feb. 2016, when the infection moved rapidly through regions occupied by *Aedes* mosquitos in the Americas, WHO declared that Zika virus infection constituted a Public Health Emergency of International Concern (PHEIC).

Zika virus infection was reported from 20 countries and territories across Americas, and an outbreak numbering thousands of cases was also identified in Cabo Verde in western Africa. In regions with no mosquito vectors, concern of spread of infection was speculated to be through travelers who sexually transmit the disease to partners who have never been to places where the virus is endemic.[2]

COMMON FEATURES OF EPIDEMICS

According to previous work on Ebola virus disease epidemic by Qureshi et al.,[3] there are several factors can begin an epidemic including the following:

1. Disasters (e.g., wars, famine, floods, and earthquakes)
2. Temporary population settlements
3. Preexisting diseases in the population
4. Ecological changes like floods and cyclones
5. Resistance potential of the host (i.e., nutritional and immunization status of the host)
6. Damage to public utility and interruption of public health services

Qureshi et al. mentioned that there are three patterns of disease continuity:

1. Saw tooth pattern
2. Tooth necklace pattern
3. Tooth eruption pattern

Saw Tooth Pattern

This pattern represents situations with an intermittent outbreak of a disease that recedes in intensity, but the disease is not eradicated from the population. The smallpox epidemics in Africa during 1920s through the 1950s would be an example of such a pattern.

Tooth Necklace Pattern

This pattern constitutes situations where the disease is eradicated from the population, but pathogen species is kept alive under controlled circumstances for preparation of vaccines and biological studies. While the escape of pathogen from confinements of laboratories has been the subject of numerous conspiracy theories, vaccination with live attenuated viruses is more likely to be the string to maintain the continuity.

Tooth Eruption Pattern

This pattern constitutes situations where, like the tooth hidden within the gums and emerging independent of other teeth, the pathogen emerges and is exterminated without any relation to previous occurrences. The Zika virus is one of the pathogens following the "tooth eruption" pattern where the first human infection was reported in 1952, followed by initial spread from Asia to Pacific Island in 2007, reports about virus leading to Guillain-Barré syndrome in adults and microcephaly in babies of mothers infected with Zika virus infection during pregnancy in 2014 and 2015, respectively.

WHY EPIDEMICS DIE THEIR DEATHS?

It is generally believed that measures such as vaccination of at-risk individuals, quarantine of diseased persons, and acute and timely treatment help to control all the epidemics. However, facts do not support this conclusion. In fact, the largest epidemics, such as the Peloponnesian War Pestilence, Antonine Plague, Plague of Justinian, Black Death of the fourteenth century, and Spanish flu, came to an end without widespread use of any of these strategies.

Qureshi et al.[3] proposed three theories for spontaneous remission of epidemics, which are listed below:

1. There are two types of people within the exposed population: some more vulnerable and some more resistant. The people who may be resistant to the disease may be so because of previous exposure to viruses with similar structures, resulting in the development of immune responses that are adequate for multiple pathogens. They might also be resistant due to superior health, including age, nutritional, and occupational advantages. The virus might eventually be faced with a high proportion of population that is partially or completely resistant to the infection.
2. Changing environment within habitats that are not conducive to the survival or propagation of viruses or other pathogens. Weather changes, including temperature and humidity fluctuations, may significantly influence the survival or propagation of a virus outside the body. Elimination of reservoirs that carry pathogens including animals, insects, food, or water, by chance or design, may disrupt the cycle of propagation. Such elimination of infection is less likely to occur within an epidemic because of diverse factors and geographical areas involved.
3. The most likely explanation is the "Sand Filter Theory," a term coined by Adnan I. Qureshi, MD. This theory reflects the similarity between retention of particulate matter during filtration based on density of sand particles, which can be compared to pathogens within a population based on population density. Most epidemics are composed of diseases that require close contact between diseased and healthy individuals for continued propagation of pathogens. Unlike natural disasters, such as hurricanes, floods, volcanoes, and changes in climate that exist independent of population density, epidemics depend upon population density, a feature shared with reproduction rates, migrations, and predation. After population density reduces below a critical limit, such contact may not allow for continued propagation of pathogens.

POTENTIAL MECHANISMS OF ZIKA VIRUS DISEASE EPIDEMIC EMERGENCE SINCE 2013

The sudden and dramatic emergence of Zika virus disease into a human-mosquito-human transmission cycle leading to major outbreaks since 2007 brings up an important question: How and why did the virus emerge and where and when it did cause infection? Some of the possible hypothesis explaining this emergence are listed below.

1. **Adaptive evolution for mosquito transmission**
   The Zika virus underwent an adaptation which resulted in more efficient transmission by *Aedes aegypti* and other closely related mosquitoes *Aedes hensilii* incriminated in Yap or *Aedes polynesiensis* suspected as a vector in French Polynesia. Polygenetic analysis suggests that this adaptive change might have occurred in Southeast Asia or the South Pacific. This assumption is supported by established presence of Zika virus disease in Southeast Asia since at least 1966, including its isolation from *Aedes aegypti* in Malaysia, but without evidence for major urban epidemics. An earlier example supporting this hypothesis is vector adaptive evolution of the Indian Ocean Lineage of Chikungunya virus for enhanced transmission by *Aedes albopictus* through mutations in envelope glycoprotein genes. This resulted in enhanced infection of the midgut epithelial cells. Validity of this hypothesis can be tested by comparing the infectivity of older Asian lineage Zika virus strains with recent isolates followed by testing effects of these mutations by reverse transcriptase approach.[4]

2. **Adaptive evolution for human viremia**
   Asian Zika virus lineage might have adapted recently to cause higher level of viremia in humans, leading to more efficient mosquito infection and a higher level of transmission and spread. Higher levels of viremia could lead to enhanced trans-placental transmission leading to microcephaly. Zika virus gene sequence studies suggest an increased use of human type codons in recent isolates of virus genomes which may support this hypothesis. However, potential links between human codon usage and enhanced human infection will require studies with human cells or animal models. This hypothesis is difficult to test because even nonhuman primates may not respond to Zika virus infection in exactly the same way as humans.[4]

3. **Herd immunity from endemic exposure in Africa and Asia limits outbreak magnitude**
   Herd immunity is a term used to represent immunity to a particular pathogen in the population at large formed predominantly by individuals without any clinical manifestations of the disease. Enzootic and/or endemic Zika virus exposure in Asia results in relatively stable levels of herd immunity, which decreases the risk for recognized outbreaks. Limited human seroprevalence studies in a few Asia sites support this hypothesis. Recent studies about longitudinal population immunity to Chikungunya virus infection in the Philippines suggest cyclic endemic circulations that maintain herd

immunity in the range of approximately 20–50% of individuals in the population. Such high rates of immunity may reduce the risk of major epidemics in the population. In Asia, further monitoring is needed to estimate levels of Zika virus disease transmission and seroprevalence. Logistically difficult studies need to be done to determine if an enzootic cycle exists in nonhuman primates or other vertebrates mammals. Stable endemic circulation of Zika virus in Asia could result in lack of recognition of Zika virus infections because most are clinically indistinguishable from Dengue virus infections and lack or appropriate use of Zika virus infection diagnostic tests make diagnosis more difficult (under diagnosis). Higher levels of herd immunity by the time women reach child bearing age would result in lower incidence of microcephaly than in epidemic regions of Americas. Low incidences of microcephaly in endemic regions would most likely go unnoticed where there are not enough cases to raise the suspicion of cause in the presence of other established causative etiologies such as Cytomegalovirus, Herpesviruses, Rubella, Toxoplasmosis, or toxic exposures.[4]

4. **Stochastic introductions into the Pacific Regions and Americas**
   Recent Zika virus disease outbreak into naïve populations in the South Pacific regions resulted by chance introduction of the virus where competent vectors mediated sufficient amplification to raise the risk of transport to the Americas, which is also occupied by naïve populations. In recent decades, increased air travel has enhanced the risk of introducing the virus into virus naive areas, and athletic competitions in Brazil have brought travelers from South Pacific region around the time the full magnitude of Zika virus spread was discovered.[4]

GLOBAL DISTRIBUTION OF DENGUE, ZIKA, AND CHIKUNGUNYA VIRUS INFECTION[5]

How Zika, Dengue and Chikungunya viruses spread?

Zika virus spread to humans through the bite of *Aedes* species mosquito (*Aedes aegypti* and *Aedes Albopictus*) which are also vectors for Dengue and Chikungunya virus transmission. These mosquitos lay eggs in standing water in buckets, bowls, animal dishes, flower pots, and vases. They are aggressive bitters during the daytime but can also bite at night. Geographic distribution of *Aedes aegypti* and *Aedes albopictus* mosquitoes in United States is illustrated Fig. 6.1.[5]

Other modes of transmission of Zika virus infection are illustrated in Fig. 6.2.

As we know, all three viruses (Zika, Dengue, and Chikungunya) spread through bite of *Aedes* species mosquito (*Aedes aegypti* and *Aedes Albopictus*) but differ in disease presentation and severity. Dengue Fever virus infection causes high fever, severe headaches, and joint pain, which can progress to hemorrhagic fever in which patient experiences more bleeding and persistent vomiting. If left untreated, Dengue Fever can progress to hemodynamic shock and death. Zika virus infection can lead to mild fever, rash, joint pain,

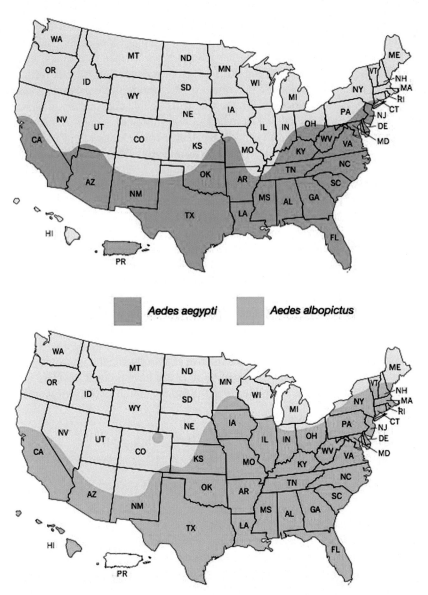

FIG. 6.1   In Apr. 2015, the Center for Disease Control and Prevention updated its vector surveillance maps to depict that both *Aedes aegypti* and *Aedes albopictus* are now believed to inhabit a wider range of distribution in the United States. Part figure (A) illustrates the expansion of territory covered by *Aedes aegypti*; Part figure (B) illustrates the expansion of *Aedes albopictus*. Both mosquitos have spread significantly to the north and west regions. Although the significance of this expansion in epidemiologic terms is unclear, it may place a greater proportion of the population at risk for exposure to emerging arboviruses such as Zika virus, particularly during warmer months.[5]

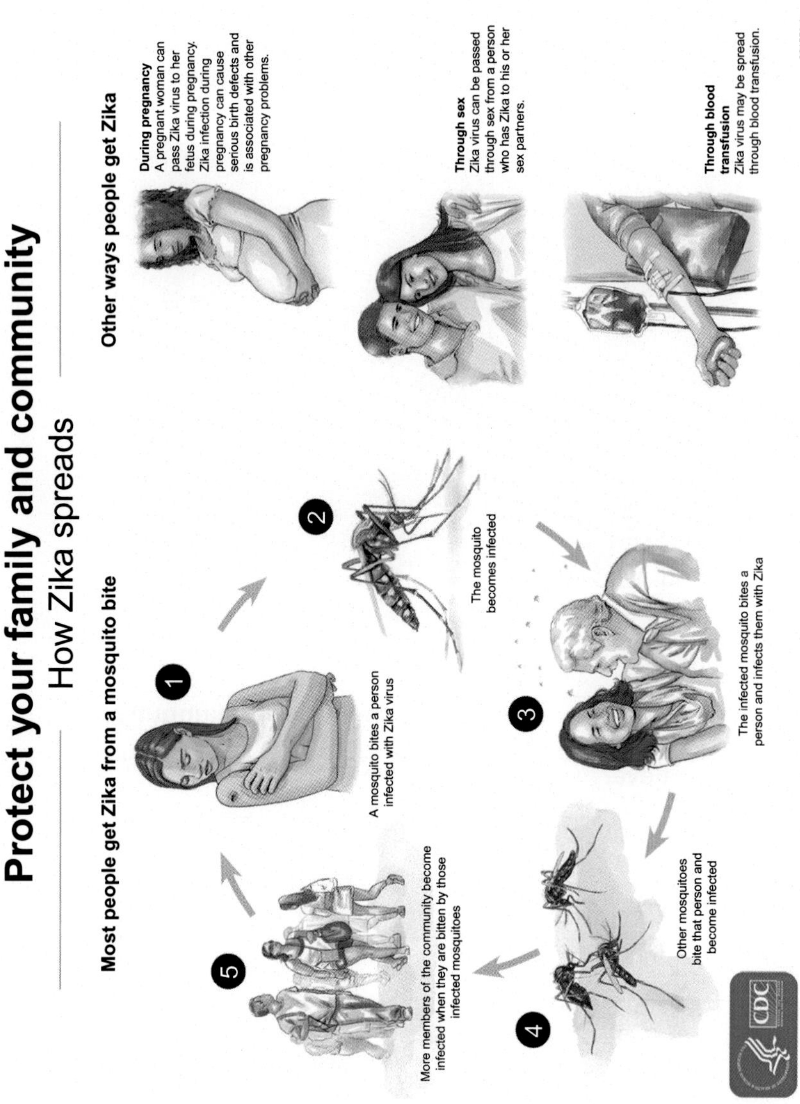

Protect your family and community
How Zika spreads

Most people get Zika from a mosquito bite

1 A mosquito bites a person infected with Zika virus

2 The mosquito becomes infected

3 The infected mosquito bites a person and infects them with Zika

4 Other mosquitoes bite that person and become infected

5 More members of the community become infected when they are bitten by those infected mosquitoes

Other ways people get Zika

During pregnancy
A pregnant woman can pass Zika virus to her fetus during pregnancy. Zika infection during pregnancy can cause serious birth defects and is associated with other pregnancy problems.

Through sex
Zika virus can be passed through sex from a person who has Zika to his or her sex partners.

Through blood transfusion
Zika virus may be spread through blood transfusion.

264550-A

CDC

FIG. 6.2 How Zika Spreads?[6]

and red eyes. Most of the infected people have no symptoms. If symptoms do occur, they occur 3–7 days after mosquito bite. Recently in February 2016, reports of Zika virus infection leading to microcephaly in babies whose mothers were infected during pregnancies were made public. Chikungunya virus causes acute onset fever, severe joint pain, headache, muscle pain, and joint swelling or rash.[7] Though all the three viruses have the same vector (mosquito) for transmission, they have a different pattern of disease presentation. It is most likely due to the viruses' different genetic makeups and virulence to different cell types (Figs.6.3 and 6.4).

FIG. 6.3 Microcephaly in babies of mothers who got infected with Zika virus disease during their pregnancies.[7]

Features	Zika	Dengue	Chikungunya
Fever	++	+++	+++
Rash	+++	+	++
Conjunctivitis	++	-	-
Arthralgia	++	+	+++
Myalgia	+	++	+
Headache	+	+	++
Hemorrhage	-	++	-
Stroke	-	+	-

FIG. 6.4 Zika virus infection compared with Dengue and Chikungunya viruses infections.[8]

TREATMENTS FOR ZIKA, DENGUE, AND CHIKUNGUNYA VIRUSES INFECTIONS

There is no specific treatment for the Zika, Dengue, and Chikungunya viruses infection. Management consists of rest and symptomatic treatment, including drinking fluids to prevent dehydration, and administration of acetaminophen to relieve fever and pain.

PREVENTION

Prevention from above-mentioned viruses-related infection include following steps.

1. **Vaccines**
   There are four vaccines which have been developed against four different filoviruses using four different approaches, (1) Yellow Fever virus (17D strain) vaccine is a live attenuated vaccine being used in human since 1937, (2) Japanese encephalitis virus vaccine is both inactivated-virus and live attenuated vaccine (14-14-2 strain), (3) tick-borne encephalitis virus vaccine is inactivated-virus vaccine, and (4) a chimeric virus vaccine which contains structural membrane and envelope genes from Dengue virus has been recently approved for clinical use for dengue virus. The above-mentioned approaches can be used to develop a vaccine for Zika virus. In addition, subunit vaccines containing Zika virus proteins, deoxyribonucleic acid vaccines expressing viral proteins, and other viral vectors expressing viral antigens can be explored, but each vaccine has its pros and cons (i.e., live attenuated can trigger more robust humeral and cellular immune response for better protection, but subunit vaccines have better safety and shorter development time). Therefore, to develop an effective vaccine for Zika virus, supplementary approaches should be explored simultaneously.[4]

2. **Therapeutics approaches**
   No clinically approved therapy is available for any Flavivirus infection. Over the past 10 years, significant effort has been put toward Dengue virus drug discovery. Due to the similarity between Dengue virus and Zika virus, knowledge gained from dengue drug discovery experience could be applied for developing effective therapy and possibly inhibitors active against both Zika virus and Dengue virus could be found. However, caution should be taken, as the biology for two viruses could be very different. Understanding disease biology is essential for therapeutic development for any pathogen. It is important to understand where in the body Zika virus replicates during course of infection and this will guide pharmacokinetics in terms of where the inhibitors should be distributed with perhaps selective penetration and accumulation.

   Duration of viremia determines the therapeutic window (time period during which treatment is most effective) after exposure to virus. To prevent microcephaly and other fetal complications from maternal Zika virus infection, the therapeutic agent should be able to cross the blood brain barrier to prevent viral replication in fetal brains, and should also be able to provide high systemic exposure to prevent replication in other systemic organs. To develop effective therapy and vaccines against Zika virus infection, animal models that mimic human diseases of Zika virus infection are urgently needed. Since the major target population for Zika virus infection is humans, it will take a little extra time to develop the effective therapy. Once the effective compound is developed, it can be used for prophylaxis for travelers, family members of infected individuals, and the general population at risk during the epidemic.[4]

3. Mosquito protection

Individuals in geographic areas with the risk for transmission should take personnel and environmental measures to avoid mosquito bites. Personnel measures include wearing long sleeve and pants, using insect repellent, and staying indoors as feasible. Persons infected should avoid mosquito bites for at least one week to avoid spread of infections to healthy individuals. Similarly, persons who travelled to a high-risk area should avoid mosquito bites for 3 weeks after returning to nonendemic areas. Environmental measures include identification and elimination of potential mosquito breeding sites of standing waters outdoors (flower pots, buckets, bottles, and jars) and in domestic water tanks.[9]

In addition, several new technologies have shown promising results in reducing mosquito populations. One approach involves releasing genetically modified male mosquitoes that express dominant lethal gene at larval stage, resulting in death of all offspring of wild female mosquitoes, leaving no risk for persistence of transgene in nature. This approach had a tremendous success in reducing population of *A. aegypti* when applied on a small scale, but logistical, technical, and financial challenges of scaling up this approach to large tropical cities where Zika virus and other arboviruses exists have not been addressed.[4]

Another approach for reducing transmission of Dengue and potentially Zika viruses is through releasing *Aedes aegypti* infected with endosymbiotic Wolbachia bacteria in natural populations. This approach can suppress the viral transmission by interfering with viral replication in mosquitos. This approach is being tested in Dengue virus infection endemic tropical regions to look for the evidence that disease incidence can be reduced. Potential limitations of this approach include (1) the need to release Wolbachia-infected mosquitoes over wide geographic ranges to overcome the limited *Aedes aegypti* flight range and (2) the possibility of arboviruses developing resistance mechanisms to overcome the inhibitory effects of Wolbachia.[4]

Finally, the most inexpensive and maintenance-free method is using lethal trap. An autocidal gravid ovitrap has been shown to reduce the mosquito population by 53 to 70% using 3–4 traps per home in 81% of houses in two isolated urban areas of Puerto Rico. Many other designs are being tested. These simple traps may be very useful when combined with source reduction and adulticide application in regions at risk for Zika virus transmission by *Aedes aegypti*.[4]

4. Sexual transmission

Sexual transmission to partners of returning male travelers who were infected with Zika virus abroad has been reported. In one case, sexual intercourse occurred before the onset of symptoms, whereas in other cases it occurred during the development of symptoms or shortly thereafter. The risk factors for and duration of risk for sexual transmission have not been determined so far. Replicative viral particles, as well as viral ribonucleic acid have been identified in sperm and viral ribonucleic acid has been detected up to 62 days after symptoms onset.[10]

DO WE NEED TO QUARANTINE ZIKA VIRUS-INFECTED PATIENTS?

Quarantine and isolation are both public health tools that involve physical separation and confinement of infected individuals to prevent disease spread and for protecting the general public health. Isolation is used for symptomatic individuals, while quarantine is for asymptomatic individuals. The decision for quarantine should be based on best available evidence rather than being driven by fear or political motivation. In general, to order a quarantine, the disease in question must be transmitted from person to person and it must occur before symptom onset. Once symptoms occur, a person would be isolated rather quarantined. The disease must also have high mortality and morbidity rate.

Zika virus infection does not meet the above-mentioned criteria as 80% of individuals incubating the disease are asymptomatic and no diagnostic test is yet available to test the infected persons rapidly. Therefore, quarantine for asymptomatic individuals would be virtually impossible. The most effective approach would be eliminating exposure to potentially infected mosquitoes and enhancing efforts to control and eradicate Zika virus carrying mosquitoes. Education of political decision makers, health care providers, and the public, especially pregnant women or ones who are planning pregnancy, about the ways to protect themselves is important. We advocate for outcome-based research, and scientific inquiry in protecting public healthcare.[11]

CONSPIRACY THEORIES ABOUT ZIKA VIRUS DISEASE PANDEMIC IN AMERICAS

The possible association of Zika virus infection with microcephaly in fetuses led to number of conspiracy theories. First, a group of Argentinian doctors claimed that a larvicide not the mosquito-borne Zika virus was the cause of the microcephaly cases. The Brazilian government clarified that microcephaly cases were also in regions where larvicide, pyriproxyfen, was not in its drinking water. Furthermore, the government was asserted that these larvicides were approved by World Health Organization and there was no scientific proof of its association with microcephaly. Second, another group blames Oxitec Company for releasing genetically modified mosquitoes in Brazil in 2011 for controlling spread of Dengue fever and believes it was behind the current epidemic of Zika virus disease. Brazilian health officials and scientists refuted this claim and emphasized that genetically modified mosquitoes reduced the mosquito population in the affected areas. Third, another common misconception is that vaccines for Chicken pox and Rubella virus were responsible for surge in microcephaly cases.[12] Due to a lack of knowledge about Zika virus infection, an aura of a mystery has developed about its origin, spread, and consequences.

REFERENCES

1. Torrey T. What Is the Difference Between an Epidemic and a Pandemic?, 2016, Retrieved on January 27, 2017 from https://www.verywell.com/.
2. Kindhauser MK, Allen T, Frank V, Santhana RS, Dye C. Zika: the origin and spread of a mosquito-borne virus. *Bull World Health Organ* 2016;**94**. 675-86c.
2a. Kindhauser MK, Allen T, Frank V, Santhana RS, Dye C. Zika: the origin and spread of a mosquito-borne virus. *Bull World Health Organ* 2016;**94**(9):675C–686C..
3. Qureshi AI. *Ebola virus disease: from origin to outbreak.* Amsterdam: Elsevier; 2016.
4. Weaver SC, Costa F, Garcia-Blanco MA, et al. Zika virus: History, emergence, biology, and prospects for control. *Antivir Res* 2016;**130**:69–80.
5. Patterson J, Sammon M, Garg M. Dengue, Zika and Chikungunya: emerging Arboviruses in the New World. *West J Emerg Med* 2016;**17**:671–9.
6. How Zika Spreads?. 2016. (Accessed May 06, 2017, at https://www.cdc.gov/zika/pdfs/zika-transmission-infographic.pdf.).
7. DISTRICT S-YMVC. Chikungunya, Dengue Fever, Yellow Fever and Zika, 2016.
8. Service ARPH. Dengue Fever, Zika & Chikungunya, 2016.
9. UpToDate. Zika virus infection: An overview, 2017.
10. Petersen LR, Jamieson DJ, Honein MA. Zika virus. *N Engl J Med* 2016;**375**:294–5.
11. Koenig KL. Quarantine for Zika Virus? Where is the Science? *Disaster Med Public Health Prep* 2016;**10**:704–6.
12. Al-Qahtani AA, Nazir N, Al-Anazi MR, Rubino S, Al-Ahdal MN. Zika virus: a new pandemic threat. *J Infect Dev Ctries* 2016;**10**:201–7.

Chapter 7

Viral Structure and Genetics

Zika virus, as the nomenclature implies, is a virus, which means that it unlike most other forms of "life", is not made up of cells. Most scientists are opponents of considering viruses to be a lifeforms, while others are proponents; however, this is a matter of definition and opinion. Viruses require a host to replicate (or reproduce). The host is usually made up of cells, and the virus utilizes the "machinery" already present inside of a cell to undergo reproduction. Although an abstract concept viruses certainly fall somewhere in the spectrum of nonliving and life.

Viruses, in general, exist as free units in nature. Viruses are generally made up of two to three major parts: (1) genetic material, known as deoxyribonucleic acids, and ribonucleic acids [Fig. 7.1, black]; (2) a coat of protein (the capsid) which surrounds the genetic material) [Fig. 7.1, red]; and (3) the envelope of fat molecules (lipids) that surround the genetic material and its capsid) [Fig. 7.1, yellow].

Each of these parts has its own function. The genetic material can be thought of as molecular instructions that ultimately allow a virus to reproduce. A good analogy to explain this is a recipe book from a library. Imagine that you are trying to make brownies for the first time. You know that your local library has a great recipe book to make them. So you go to the library, find the book, and look at the recipe, but realize that it is impossible to remember all the steps. So you know that for you to produce these brownies as home, you will probably have to photocopy the recipe and take that copy home with you. One you get home, you look at your photocopy, follow the outlined steps, and make the brownies. In this story, deoxyribonucleic acids can be thought of as the recipe book at the library. Ribonucleic acids can be thought of as the photocopy of the recipe, and the brownies can be thought of the components that assemble to create a reproduced or replicated virus. The labor that you put in to walk to the library, make the photocopy, and bake the brownies can be thought of as the host cell and it "machinery" that the virus requires to replicate. The process by which the deoxyribonucleic acids (the recipe book) is "photocopied" is formally called "transcription." The process by which the RNA (the photocopy) is used to make the viral components (the brownies) is formally called "translation."

With that background on general viral biology, a more detailed look at the viral structure and genetics of the Zika virus itself is warranted. The genetic

Zika Virus Disease. https://doi.org/10.1016/B978-0-12-812365-2.00008-1

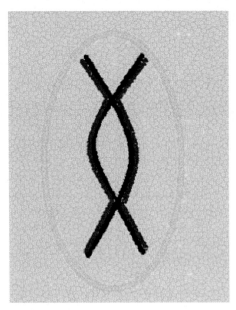

FIG. 7.1 Schematic of a typical virus structure.

material of the Zika virus is composed of ribonucleic acids, so the black lines in Fig. 7.1 of the Zika virus would be ribonucleic acids. Also, the Zika virus contains the protein coat (red) and a lipid envelope (yellow) as well. Once the Zika virus is introduced to a host cell, which was likely introduced to the human by either mosquito (bite), placental (mother to fetus), or sexual (human to human), the virus fuses with the cell to introduce its genetic material and capsid to the intracellular (inside of the cell) environment (Fig. 7.2).

Furthermore, since there is no deoxyribonucleic acids component to the Zika virus in our analogy, the step of going to the library to obtain the photocopy from the original recipe book does not occur. Therefore, one can think of the Zika virus as "the photocopy." Once the ribonucleic acids and its protein coat enter the cell, the coat is removed (Fig. 7.3).

The ribonucleic acids (photocopy of recipe) is then ready to undergo translation (being made into brownies). This is achieved by attaching to a ribosome (Fig. 7.4), which is always normally present and floating freely within the intracellular compartment.

The ribosome can be thought of as the "major player" of the machinery within a cell that a virus uses to ultimately make copies of itself. The ribonucleic acids basically runs through the ribosome, and while its being ran through, there is simultaneous manufacturing of the viral components (Fig. 7.5).

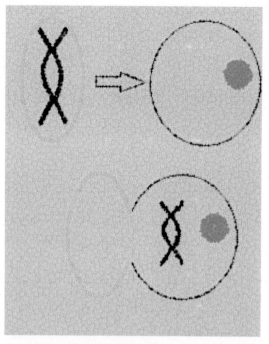

FIG. 7.2 Schematic of cellular infection by Zika virus.

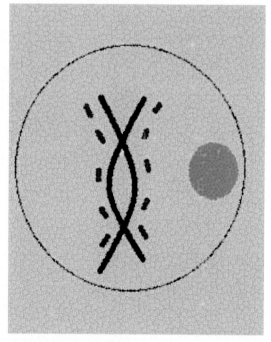

FIG. 7.3 Removal of protein coat from RNA.

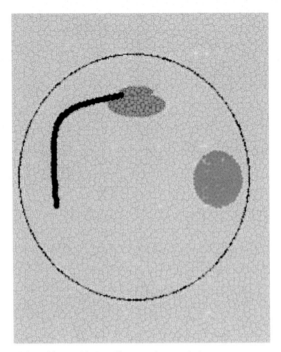

FIG. 7.4   Ribonucleic acids attaching to ribosome for translation.

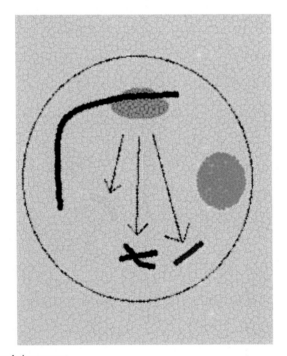

FIG. 7.5   Translation process.

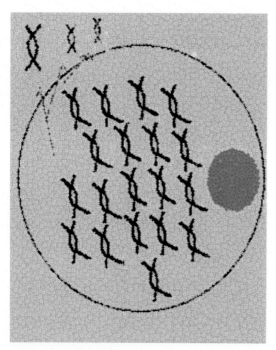

FIG. 7.6 Viral reproduction and cellular exit.

The viral components then assemble. After assembly, the new viruses leave the cell to infect other nearby cells, in which this process repeats itself all over again (Fig. 7.6).

Chapter 8

Clinical Manifestations and Laboratory Diagnosis of Zika Virus Disease

Chapter Outline

PATHOGENESIS OF DISEASE

Zika virus, is a neurotropic virus.[1] Neurotropism is based on a cell surface protein called AXL protein, which acts as a receptor for Zika virus entry into the cell.[2] AXL is expressed in high proportions in the developing fetal human cortex.[3] AXL is also expressed in cutaneous fibroblasts and epidermal keratinocytes. The high expression of AXL within the human skin at the site of inoculation represents the first site of viral replication.[4] From the skin, the virus spreads to the draining lymph node, where it is amplified by replication of virions in local tissue cells and the regional lymph nodes, resulting in viremia and hematogenous dissemination.[5] Recently, the Asian Zika virus lineage may have adapted to generate higher viremia in humans, which explains the recent spread of infection by mosquitoes. Higher viremia load could enhance transplacental transmission, which explains the emergence of microcephaly.[6] Zika virus infect human neural progenitor cells, which are multipotent cells that generate the neurons and glial cells during embryogenesis—subsequently leading to cell-cycle arrest, apoptosis, and inhibition of neural progenitor cells differentiation, resulting in microcephaly in fetuses infected with Zika virus.[7] The viral infection triggers apoptosis of infected cells in order to increase the release of infectious viral particles.[8]

CLINICAL PRESENTATION

The majority of infections are asymptomatic (80%), and severe complications, such as Guillain-Barré syndrome, are rare.[9] However, during the period from March 2015 to February 2016, a greater than 20-fold increase in microcephaly cases was observed in Brazilian newborns compared with previous

Zika Virus Disease. https://doi.org/10.1016/B978-0-12-812365-2.00009-3

103

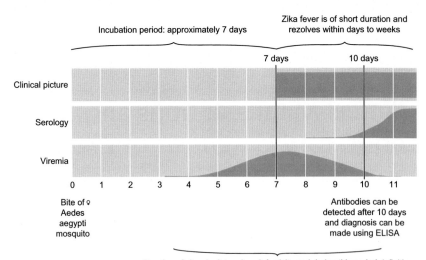

FIG. 8.1   Diagram for illustration of ZIKV infection time intervals. Time intervals are approximate and do not indicate the actual duration. *RT-PCR*, reverse transcription polymerase chain reaction; *ELISA*, enzyme-linked immunosorbent assay.

years and a relationship between microcephaly and Zika virus infection during pregnancy was observed.[10–13]

Zika virus disease—The typical presentation of infection by Zika virus is in the form of acute febrile illness of short duration where fever resolves within days to weeks (Fig. 8.1).[14,15] Zika virus is one of the Flaviviruses and the initial presentation can resembles a mild form of Dengue fever.[16] That is why all diagnoses of "atypical Dengue" in travelers returning from areas where Zika virus infection is endemic should be carefully investigated.[17] The incubation period for Zika virus infection is 3–12 days (Fig. 8.1). Symptoms include low-grade fever, maculopapular pruritic rash, arthritis or arthralgia, conjunctivitis, myalgia, and headache.[18]

Inflammatory-autoimmune neurological syndromes—Many studies showed a relationship between Guillain-Barré syndrome and Zika virus infection outbreaks.[10,19,20] The incidence of Guillain-Barré syndrome is estimated to occur in approximately 2 out of every 10,000 cases of Zika virus infected individuals.[21] Guillain-Barré syndrome includes a wide spectrum of variants like Miller Fisher syndrome and Bickerstaff's brain-stem encephalitis.[22] These long term sequel are described in later chapters.

Teratogenicity—The classical group of teratogenic pathogens associated with congenital manifestations is summarized by the acronym TORCH (Toxoplasmosis, others [Syphilis and Human immunodeficiency virus], Rubella, Cytomegalovirus, and Herpes Simplex virus). Zika Virus now joins the list of TORCH infections.[23] Fetal complications include microcephaly, chorioretinal atrophy, placental insufficiency, fetal growth restriction, and fetal death.[24–26] Microcephaly is a clinical finding of a small head size for gestational age and is indicative of an underlying problem with the growth of the brain. Microcephaly can occur as a result of the poor growth of the fetal skull that follows the growth restriction of fetal brain tissue. Microcephaly is defined as a postnatal occipito-frontal circumference below the negative second standard deviation (i.e., below the 2.3 percentile).[27, 28]

Hemorrhagic complications—Laboratory tests are generally in the normal range. However, in some cases, leucopenia and mild thrombocytopenia have been described and mild hemorrhagic symptoms (e.g., petechiae) were reported.[5]

Death—In addition to neonates who died in the first 24 hours, three deaths due to Zika virus infection were reported in late November 2015 in Brazil (two of them were adults)[17]. One of the reasons for the deaths in adults is that Zika virus infection results in the development of immune thrombocytopenic purpura which may cause a fatal hemorrhagic disorders.[29] Another reason is death from severe respiratory muscle paralysis and respiratory failure among Guillan-Barre syndrome patients. One of the deaths was attributed to septic shock in Guillan-Barre syndrome patient.[30] Finally, development of acute meningoencephalitis due to Zika virus infection can be fatal especially in an immunocompromised patient.[31]

DIAGNOSIS OF ZIKA VIRUS INFECTION

Many laboratory studies have been used to diagnose Zika virus infection. It can be categorized into virus isolation, serological studies and molecular studies.

Virus isolation and culture—Isolation of Zika virus was first done in 1947 from serum samples taken from a rhesus monkey in Zika forest in Uganda and the culture medium was mouse brain (see Table 8.1). Subsequent isolation and culture methods used include inoculation of chicken embryo yolk sacs, and cell cultures. Zika virus has been successfully cultured from human blood, semen, and urine specimens.

Molecular studies—Definitive diagnosis is based on detection of Zika virus ribonucleic acid in blood or other body fluids by polymerase chain reaction.[74] Since the duration of viremia is transient (expected to last approximately 1 week although not yet well known), the detection of viral ribonucleic acid in body specimens by polymerase chain reaction is most successful within the first days of clinical illness.[75] However, it is to be mentioned that there are reports that Zika virus ribonucleic acid can be detected in urine up to 20 days from the onset of clinical symptoms, thus increasing the duration during which the definitive diagnosis can be made.[76–78]

Serological studies—Since its isolation in 1947, different serological tests were used to diagnose Zika virus infection from mouse protection test in the 1950s to neutralization test and hamagglutination tests, complement fixation and currently enzyme-linked immunosorbent assay has become the leading test (Table 8.2).

TABLE 8.1 The Evolution of Diagnostic Tests Used for Identification of Zika Virus Infection

Reference	Year	Country or Region	Test Used
Dick et al.[32]	1947	Uganda	Virus isolated from a rhesus monkey
MacNamara et al.[33]	1951	Nigeria	Mouse protection test
Smithburn[34]	1952	Uganda and Tanzania	Neutralization test
Smithburn et al.[35]	1952	India	Neutralization test a. Two of the samples -of 196 tested- neutralized only Zika virus.
Smithburn[36]	1953	Malayan peninsula and North Borneo (today state of Sabah, Malaysia)	Neutralization test
MacNamara[37]	1953	Nigeria	In two cases by neutralization test and in the one case by isolation of the virus (the first human Zika virus isolate)
Smithburn et al.[38]	1954	Egypt	Neutralization test
MacNamara et al.[39]	1955	Nigeria	Mouse protection test
Kokernot et al.[40]	1957	Mozambique	Neutralization test
Weinbren and Williams[41]	1958	Uganda	Virus isolation from Aedes africanus mosquito
Chippaux-Hyppolite[42]	1961–1962	Central African Republic	Hemagglutination assay
Brès et al.[43]	1962	Senegal	Hemagglutination inhibition test
Chippaux-Hyppolite[44]	1963–1964	Central African Republic	Hemagglutination inhibition test
Brès[45]	1963–1964	Burkina Faso	Hemagglutination inhibition test
Brès[45], Robin et al.[46]	1963–1965	Côte d'Ivoire	Hemagglutination inhibition test
Brès[45], Pinto[47]	1964–1965	Guinea-Bissau	Hemagglutination inhibition test

TABLE 8.1 The Evolution of Diagnostic Tests Used for Identification of Zika Virus Infection—cont'd

Reference	Year	Country or Region	Test Used
Brès[45]	1964–1966	Togo	Hemagglutination inhibition test
Brès[45]; Salün and Brottes[48]	1964–1966	Cameroon	Hemagglutination inhibition test
Brès[45]	1964–1967	Mali	Hemagglutination inhibition test
Simpson[49]	1964	Uganda	First report and confirmation that Zika virus causes human disease Virus isolation by isolating the virus from patient's blood, and by infecting mice and re-isolating the virus from mice blood
Casals[50], Musso and Gubler[51], Monath et al.[52], Robin[53]	1965–1967	Nigeria	Hemagglutination inhibition test
Brès[45]	1967	Benin	Hemagglutination inhibition test
Brès[45]	1967	Gabon	Hemagglutination inhibition test
Brès[45]	1967	Liberia	Hemagglutination inhibition test
Henderson et al.[54]	1966–1967	Uganda	Hemagglutination inhibition test
Henderson et al.[54], Geser et al.[55]	1966–1967	Kenya	Hemagglutination inhibition test
Henderson et al.[54]	1966–1967	Somalia	Hemagglutination inhibition test
Brès[45]	1966–1967	Morocco	Hemagglutination inhibition test
Henderson et al.[56]	1967–1969	Uganda	Hemagglutination inhibition test

(Continued)

TABLE 8.1 The Evolution of Diagnostic Tests Used for Identification of Zika Virus Infection—cont'd

Reference	Year	Country or Region	Test Used
Henderson et al.[57]	1968	Kenya	Hemagglutination inhibition test
Fagbami et al.[58]	1969–1972	Nigeria	Neutralization test
Marchette et al.[59]	1969	Malaysia	Virus isolated from a pool of 29 *Aedes aegypti* mosquitoes
Monath et al.[60]	1970	Nigeria	Hemagglutination inhibition test
Filipe et al.[61]	1971–1972	Angola	Hemagglutination inhibition test
Renaudet et al.[62]	1972, 1975	Senegal	Hemagglutination inhibition test
Gonzales et al.[63]	1979	Central African Republic	Hemagglutination inhibition test
Adekolu-John and Fagbami.[64]	1980	Nigeria	Hemagglutination inhibition test
Rodhain et al.[65]	1984	Uganda	Hemagglutination inhibition test
Monlun et al.[66]	1988, 1990	Senegal	IgM ELISA
Wolfe et al.[67]	1996–1997	Malaysia	Neutralization test
Akoua-Koffi et al.[68]	1999	Côte d'Ivoire	IgG ELISA
Duffy et al.[69]	2007	Yap State (Federated States of Micronesia)	IgM ELISA
Fokam et al.[70]	2010	Cameroon	Hemagglutination inhibition test and complement fixation
Aubry et al.[71]	2011–2013	French Polynesia	IgG ELISA
Aubry et al.[72]	2014	French Polynesia	IgG ELISA
Babaniyi et al.[73]	2014	Zambia	IgM ELISA

ELISA, enzyme-linked immunosorbent assay.

TABLE 8.2 Definition of Categories of Case Classification

Category	Criteria
Confirmed	Zika virus ribonucleic acid is detected or if all the following are present: IgM antibody by enzyme-linked immunosorbent assay with confirmation by plaque reduction neutralization test where Zika virus $PRNT_{90}$ titer ≥ 20 and Zika virus $PRNT_{90}$ titer to Dengue virus $PRNT_{90}$ ratio ≥ 4
Probable	Lack of Zika virus ribonucleic acid but IgM antibody detection by enzyme-linked immunosorbent assay with confirmation by plaque reduction neutralization test where Zika virus $PRNT_{90}$ titer ≥ 20 but Zika virus PRNT90 titer to Dengue virus $PRNT_{90}$ ratio was < 4
Still suspected	No serologic and polymerase chain reaction evidence and no serum samples collected within 10 days after the onset of symptoms
Negative	No serologic and polymerase chain reaction evidence despite the fact that serum samples were collected within 10 days after the onset of symptoms

PRNT, plaque-reduction neutralization test.

Neutralizing Test: Neutralization is loss of virus infectivity through reaction of the virus with specific antibody. Virus and serum are mixed and then active virus is detected by reactions such as cytopathic effect or plaque formation or disease in animals. The "mouse protection test" is based on injecting a mixture of patient's serum and virus into mice. If the mice survived, then the virus was considered neutralized. If the mice died, the patient's serum did not contain the neutralizing antibodies and diagnosis of patient infection can be made.

Hemagglutination Inhibition Test: The nucleic acid of Zika virus encodes surface proteins that bind to the erythrocytes forming a lattice (hemagglutination) which accumulates irregularly in a tube or microtiter well. When hemagglutination is identified, antibodies to Zika virus are added to the setup which inhibits attachment of the virus to erythrocytes. Multiple dilutions of serum containing antibodies are tested and the highest dilution of serum (antibodies) that prevents hemagglutination is identified as the hemagglutination inhibition titer of the serum.

Complement Fixation Test: The complement fixation test is one of the major traditional tests identifying presence of specific antibodies. The complement fixes itself to antigen-antibody complex. In the complement fixation test, Sheep erythrocytes—antisheep erythrocyte antibody complex is used as an indicator where erythrocytes are used as the target cells. In the presence of specific antibodies to an infectious agent, any complement in the system is bound, leaving no residual complement for reaction with antibodies to the erythrocytes. Thus, the presence of specific antibody is indicated by the absence of hemolysis. On the other hand if the patient's serum does not contain antibodies, hemolysis will occur.

Enzyme-Linked Immunosorbent Assay: Most enzyme linked immunosorbent assays developed for the detection of Zika virus antibody consist of use of corresponding Zika virus antigen which is firmly fixed on a tube. Patient's serum is then added. In case of the presence of antibodies in patient's serum, there will be antigen-antibody complex, i.e., the antibody is immobilized. Later enzyme labelled antibody is added in the complex, which will combine with Fc portion of the antibody and get also immobilized. A washer is then applied, followed by substrate of the enzyme. The enzyme usually hydrolyses the substrate to give a color. The intensity of the color is proportional to the amount of antibody present in the patient's serum. On the other hand, if there is no Zika virus antibody, the complex will not form. Therefore, when the enzyme labelled antibody is added, it will not be fixed and thus will be washed and when the substrate is added, it will not get hydrolyzed and no color will appear. Because of strong serologic cross-reactivity of closely related flaviviruses, the sensitivity and specificity of five Zika virus serologic immunoassays was evaluated in one study. The evaluation included ELISA assays based on a conventional IgM (Euroimmun IgM) and three IgM antibody capture (MAC-ELISAs): Novatec IgM, Abcam IgM and InBios IgM. Zika Euroimmun IgM ELISA was also tested in parallel with the Euroimmun conventional IgG ELISA. The Euroimmun IgM, Euroimmun IgG and the Abcam IgM ELISAs showed a specificity of 100% for flavivirus seronegative specimens and >90% Dengue virus-positive samples. The Novatec ELISA showed a specificity of 66% for Flavivirus seronegative specimens and 70% for Dengue virus-positive specimens. InBios ELISA showed similar specificity results for Flavivirus seronegative specimens, but it showed decreased specificity for Dengue virus-positive samples. This assay incorrectly identified 40% of these samples as containing Zika virus IgM antibodies. The IgM assays of Euroimmun, Abcam, and Novatec demonstrated sensitivities of 37%, 57%, and 65%, respectively. The InBios IgM ELISA demonstrated a sensitivity of 100%.[31]

Diagnostic studies suggest that IgM antibodies appear as viremia wanes and persists for several months (Fig. 8.1).[75] The considerable cross-reactivity of flavivirus antibodies presents major challenges for the interpretation of serologic test results.[79] The plaque-reduction neutralization test, the most specific test used to differentiate antibodies of closely related viruses, can be used to help verify enzyme-linked immunosorbent assay results.[80] The number of plaques (regions of infected cells) is identified after confluent monolayer of host cells is exposed to serum sample and virus suspension. However, this test takes up to a week to perform and requires standardized reagents that often are not available.[81] In settings where plaque-reduction neutralization test is not available or the volume of testing makes plaque-reduction neutralization test impractical, specimens that have serological evidence of Zika virus exposure and no serological evidence of Dengue virus exposure may be interpreted as recent Zika virus infection. However, the diagnostic accuracy of this approach has not been established and is particularly problematic in areas in which dengue is endemic (where more than 90% of the population may have had previous exposure to Dengue virus).[75]

Criteria for Diagnosis—The diagnosis of Zika virus infection is based clinical presentation, and molecular and serological studies (Fig. 8.1).[82]A good example for multifaceted approach is the one used during Yap island outbreak in 2007. This approach included case definition and case classification. Case definition required suspected patients to have acute onset of generalized macular or popular rash, arthralgia, or conjunctivitis. Blood samples were acquired during the acute phase (i.e., within 10 days after the onset of symptoms) and during the convalescent phase (i.e., 14 days later) from suspected patients. Case classification categorized suspected patients into four categories: (1) confirmed, (2) probable, (3) still suspected, and (4) (Table 8.2).[69] As can be seen in Table 8.2, the definition of various categories of cases are based on combination of clinical criteria supplemented by laboratory tests. Table 8.1 provides a summary of laboratory based confirmation tests used in previous studies pertaining to Zika virus infection. Hemagglutination test has been one of the most frequently used test with detection of IgM antibodies against Zika virus by ELISA gaining more acceptance in recent studies.

REFERENCES

1. Ramos da Silva S, Gao SJ. Zika virus: an update on epidemiology, pathology, molecular biology, and animal model. *J Med Virol* 2016;**88**(8):1291–6.
2. Nowakowski TJ, et al. Expression analysis highlights AXL as a candidate zika virus entry receptor in neural stem cells. *Cell Stem Cell* 2016;**18**(5):591–6.
3. Tabata T, et al. Zika virus targets different primary human placental cells, suggesting two routes for vertical transmission. *Cell Host Microbe* 2016;**20**(2):155–66.
4. Ji R, et al. TAM receptors affect adult brain neurogenesis by negative regulation of microglial cell activation. *J Immunol* 2013;**191**(12):6165–77.
5. Barzon L, et al. Zika virus: from pathogenesis to disease control. *FEMS Microbiol Lett* 2016;**363**(18):1–17.
6. Faye O, et al. Molecular evolution of Zika virus during its emergence in the 20(th) century. *PLoS Negl Trop Dis* 2014;**8**(1):e2636.
7. Li C, et al. Zika virus disrupts neural progenitor development and leads to microcephaly in mice. *Cell Stem Cell* 2016;**19**(1):120–6.
8. Briant L, et al. Role of skin immune cells on the host susceptibility to mosquito-borne viruses. *Virology* 2014;**464–465**:26–32.
9. Lessler J, et al. Assessing the global threat from Zika virus. *Science* 2016;**353**(6300). p. aaf8160.
10. Broutet N, et al. Zika virus as a cause of neurologic disorders. *N Engl J Med* 2016;**374**(16): 1506–9.
11. Franca GV, et al. Congenital Zika virus syndrome in Brazil: a case series of the first 1501 livebirths with complete investigation. *Lancet* 2016;**388**(10047):891–7.
12. Kleber de Oliveira W, et al. Increase in reported prevalence of microcephaly in infants born to women living in areas with confirmed Zika virus transmission during the first trimester of pregnancy—Brazil. *MMWR Morb Mortal Wkly Rep, 2016* 2015;**65**(9):242–7.

13. Schuler-Faccini L, et al. Possible association between Zika virus infection and microcephaly—Brazil, 2015. *MMWR Morb Mortal Wkly Rep* 2016;**65**(3):59–62.

14. Brasil P, et al. Zika virus outbreak in rio de janeiro, Brazil: Clinical characterization, epidemiological and virological aspects. *PLoS Negl Trop Dis* 2016;**10**(4):e0004636.

15. Simpson DI. Zika virus infection in man. *Trans R Soc Trop Med Hyg* 1964;**58**:335–8.

16. Weaver SC, et al. Zika virus: History, emergence, biology, and prospects for control. *Antiviral Res* 2016;**130**:69–80.

17. Musso D, Gubler DJ. Zika virus. *Clin Microbiol Rev* 2016;**29**(3):487–524.

18. Brasil P, et al. Zika virus infection in pregnant women in Rio de Janeiro. *N Engl J Med* 2016;**375**(24):2321–34.

19. Krauer F, Riesen M. Zika virus infection as a cause of congenital brain abnormalities and Guillain-Barre syndrome: systematic review. *PLoS Med* 2017;**14**(1):e1002203.

20. Kandel N, et al. Detecting Guillain-Barre syndrome caused by Zika virus using systems developed for polio surveillance. *Bull World Health Organ* 2016;**94**(9):705–8.

21. Richard V, Paoaafaite T, Cao-Lormeau VM. Vector competence of French Polynesian Aedes aegypti and Aedes polynesiensis for Zika virus. *PLoS Negl Trop Dis* 2016;**10**(9):e0005024.

22. Lucchese G, Kanduc D. Zika virus and autoimmunity: From microcephaly to Guillain-Barre syndrome, and beyond. *Autoimmun Rev* 2016;**15**(8):801–8.

23. Coyne CB, Lazear HM. Zika virus—reigniting the TORCH. *Nat Rev Microbiol* 2016;**14**(11):707–15.

24. Mayor S. Zika infection in pregnancy is linked to range of fetal abnormalities, data indicate. *BMJ* 2016;**352**:i1362.

25. Oliveira Melo AS, et al. Zika virus intrauterine infection causes fetal brain abnormality and microcephaly: tip of the iceberg? *Ultrasound Obstet Gynecol* 2016;**47**(1):6–7.

26. Wiwanitkit V. Placenta, Zika virus infection and fetal brain abnormality. *Am J Reprod Immunol* 2016;**76**(2):97–8.

27. Ashwal S, et al. Practice parameter: Evaluation of the child with microcephaly (an evidence-based review): Report of the Quality Standards Subcommittee of the American Academy of Neurology and the Practice Committee of the Child Neurology Society. *Neurology* 2009;**73**(11):887–97.

28. Villar J, et al. International standards for newborn weight, length, and head circumference by gestational age and sex: the Newborn Cross-Sectional Study of the INTERGROWTH-21st Project. *Lancet* 2014;**384**(9946):857–68.

29. Azevedo RSS, Araujo MT, Martins Filho AJ, Oliveira CS, Nunes BTD, Cruz ACR, Vasconcelos PFC. Zika virus epidemic in Brazil. I. Fatal disease in adults: clinical and laboratorial aspects. *J Clin Virol* 2016;**85**:56–64.

30. Dirlikov E, Major CG, Mayshack M, Medina N, Matos D, Ryff KR, et al. Guillain-Barré syndrome during ongoing Zika virus transmission — Puerto Rico, January 1–July 31, 2016. *Morb Mortal Wkly Rep* 2016;**65**(34):910–14.

31. Safronetz D, Sloan A, Stein DR, Mendoza E, Barairo N, Ranadheera C, et al. Evaluation of 5 commercially available Zika virus immunoassays. *Emerg Infect Dis* 2017;**23**(9):1577–80.

32. Dick GW, Kitchen SF, Haddow AJ. Zika virus. I. Isolations and serological specificity. *Trans R Soc Trop Med Hyg* 1952;**46**(5):509–20.

33. MacNamara FN, Horn DW, Porterfield JS. Yellow fever and other arthropod-borne viruses; a consideration of two serological surveys made in South Western Nigeria. *Trans R Soc Trop Med Hyg* 1959;**53**(2):202–12.

34. Smithburn KC. Neutralizing antibodies against certain recently isolated viruses in the sera of human beings residing in East Africa. *J Immunol* 1952;**69**(2):223–34.

35. Smithburn KC, Kerr JA, Gatne PB. Neutralizing antibodies against certain viruses in the sera of residents of India. *J Immunol* 1954;**72**(4):248–57.

36. Smithburn KC. Neutralizing antibodies against arthropod-borne viruses in the sera of long-time residents of Malaya and Borneo. *Am J Hyg* 1954;**59**(2):157–63.

37. MacNamara FN. Zika virus: a report on three cases of human infection during an epidemic of jaundice in Nigeria. *Trans R Soc Trop Med Hyg* 1954;**48**(2):139–45.

38. Smithburn KC, Taylor RM, Rizk F, Kader A. Immunity to certain arthropod-borne viruses among indigenous residents of Egypt. *Am J Trop Med Hyg* 1954;**3**(1):9–18.

39. MacNamara FN, Horn DW, Porterfield JS. Yellow fever and other arthropod-borne viruses; a consideration of two serological surveys made in South Western Nigeria. *Trans R Soc Trop Med Hyg* 1959;**53**(2):202–12.

40. Kokernot RH, Smithburn KC, Gandara AF, McIntosh BM, Heymann CS. Neutralization tests with sera from individuals residing in Mozambique against specific viruses isolated in Africa, transmitted by arthropods. *An Inst Med Trop (Lisb)* 1960;**17**:201–30.

41. Weinbren MP, Williams MC. Zika virus: further isolations in the Zika area, and some studies on the strains isolated. *Trans R Soc Trop Med Hyg* 1958;**52**(3):263–8.

42. Chippaux-Hyppolite C. Immunologic investigation on the frequency of arbovirus in man in the Central African Republic. Preliminary note. *Bull Soc Pathol Exot Filiales* 1965;**58**(5):812–20.

43. Brès P, Lacan A, Diop I, Michel R, Peretti P, Vidal C. Les arbovirus au s'en'egal. enqu ete s'erologique. *Bull Soc Pathol Exot Filiales* 1963;**56**:384–402.

44. Chippaux-Hyppolite C, Chippaux A. Les anticorps antiamarils chez les enfants en République centrafricaine. *Bull World Health Organ* 1966;**34**(1):105–11.

45. Brès P. Données récentes apportées par les enquêtes sérologiques sur la prévalence des arbovirus en Afrique, avec référence spéciale à la fièvre jaune. *Bull World Health Organ* 1970;**43**(2):223–67.

46. Robin Y, Brès P, Lartigue JJ, Gidel R, Lefèvre M, Athawet B, et al. Les arbovirus en Cote-D'Ivoire. Enquête sérologique dans la population humaine. *Bull Soc Pathol Exot Filiales* 1968;**61**(6):833–45.

47. Pinto MR. Survey for antibodies to arboviruses in the sera of children in Portuguese Guinea. *Bull World Health Organ* 1967;**37**(1):101–8.

48. Salaün JJ, Brottes H. Les arbovirus au Cameroun: enquête sérologique. *Bull World Health Organ* 1967;**37**(3):343–61.

49. Simpson DI. Zika virus infection in man. *Trans R Soc Trop Med Hyg* 1964;**58**(4):335–8.

50. Casals J. *9th international congress for microbiology, Moscow, 1966*. Oxford: Pergamon; 2016, pp. 441–52.

51. Musso D, Gubler DJ. Zika virus. *Clin Microbiol Rev* 2016;**29**(3):487–524.

52. Monath TP, Wilson DC, Casals J. The 1970 yellow fever epidemic in Okwoga District, Benue Plateau State, Nigeria. 3. Serological responses in persons with and without pre-existing heterologous group B immunity. *Bull World Health Organ* 1973;**49**(3):235–44.

53. Robin Y. *Rapport de la 7ème conference technique de l'Organisation de Coordination et de Coopération pour la lutte contre les Grandes Endémies*. Bobo-Dioulasso: OCCGE; 1967.

54. Henderson BE, Metselaar D, Cahill K, Timms GL, Tukei PM, Williams MC. Yellow fever immunity surveys in northern Uganda and Kenya and eastern Somalia, 1966–67. *Bull World Health Organ* 1968;**38**(2):229–37.

55. Geser A, Henderson BE, Christensen S. A multipurpose serological survey in Kenya. 2. Results of arbovirus serological tests. *Bull World Health Organ* 1970;**43**(4):539–52.

56. Henderson BE, Kirya GB, Hewitt LE. Serological survey for arboviruses in Uganda, 1967–69. *Bull World Health Organ* 1970;**42**(5):797–805.

57. Henderson BE, Metselaar D, Kirya GB, Timms GL. Investigations into yellow fever virus and other arboviruses in the northern regions of Kenya. *Bull World Health Organ* 1970;**42**(5):787–95.

58. Fagbami AH, Monath TP, Fabiyi A. Dengue virus infections in Nigeria: a survey for antibodies in monkeys and humans. *Trans R Soc Trop Med Hyg* 1977;**71**(1):60–5.

59. Marchette NJ, Garcia R, Rudnick A. Isolation of Zika virus from Aedes aegypti mosquitoes in Malaysia. *Am J Trop Med Hyg* 1969;**18**(3):411–15.

60. Monath TP, Wilson DC, Casals J. The 1970 yellow fever epidemic in Okwoga District, Benue Plateau State, Nigeria. 3. Serological responses in persons with and without pre-existing heterologous group B immunity. *Bull World Health Organ* 1973;**49**(3):235–44.

61. Filipe AR, De Carvalho RG, Relvas A, Casaca V. Arbovirus studies in Angola: serological survey for antibodies to arboviruses. *Am J Trop Med Hyg* 1975;**24**(3):516–20.

62. Renaudet J, Jan C, Ridet J, Adam C, Robin Y. Enquête sérologique pour les arbovirus dans la population humaine du sénégal. *Bull Soc Pathol Exot Filiales* 1978;**71**(2):131–40.

63. Gonzalez JP, Saluzzo JF, Hervé JP, Geoffroy B. Enquête sérologique sur la prévalence des arbovirus chez l'homme en milieu forestier et périforestier de la région de la Lobaye (République centrafricaine). *Bull Soc Pathol Exot Filiales* 1979;**72**(5–6):416–23.

64. Adekolu-John EO, Fagbami AH. Arthropod-borne virus antibodies in sera of residents of Kainji Lake Basin, Nigeria 1980. *Trans R Soc Trop Med Hyg* 1983;**77**(2):149–51.

65. Rodhain F, Gonzalez JP, Mercier E, Helynck B, Larouze B, Hannoun C. Arbovirus infections and viral haemorrhagic fevers in Uganda: a serological survey in Karamoja district, 1984. *Trans R Soc Trop Med Hyg* 1989;**83**(6):851–4.

66. Monlun E, Zeller H, Le Guenno B, Traoré-Lamizana M, Hervy JP, Adam F, et al. Surveillance de la circulation des arbovirus d'intérêt médical dans la région du Sénégal oriental (1988–1991). *Bull Soc Pathol Exot* 1993;**86**:21–8.

67. Wolfe ND, Kilbourn AM, Karesh WB, Rahman HA, Bosi EJ, Cropp BC, et al. Sylvatic transmission of arboviruses among Bornean orangutans. *Am J Trop Med Hyg* 2001;**64**(5–6):310–16.

68. Akoua-Koffi C, Diarrassouba S, Bénié VB, Ngbichi JM, Bozoua T, Bosson A, et al. Investigation autour d'un cas mortel de fièvre jaune en Côte d'Ivoire en 1999. *Bull Soc Pathol Exot* 2001;**94**(3):227–30.

69. Duffy MR, Chen T-H, Hancock WT, Powers AM, Kool JL, Lanciotti RS, et al. Zika virus outbreak on Yap Island, Federated States of Micronesia. *N Engl J Med* 2009;**360**(24):2536–43.

70. Fokam EB, Levai LD, Guzman H, Amelia PA, Titanji VP, Tesh RB, Weaver SC. Silent circulation of arboviruses in Cameroon. *East Afr Med J* 2010;**87**:262–8.

71. Aubry M, Finke J, Teissier A, Roche C, Broult J, Paulous S, Desprès P, Cao-Lormeau VM, Musso D. 2015. Seroprevalence of arboviruses among blood donors in French Polynesia, 2011–2013. *Int J Infect Dis* 2015;**41**:11–12.

72. Aubry A, Teissier A, Roche C, Teururai1 S, Paulous S, Desprès P, Musso D, Mallet HP, Merceron S, Huart M, Sicard S, Deparis X, Cao-Lormeau VM. *Serosurvey of dengue, Zika and other mosquito-borne viruses in French Polynesia, abstr 765*. In: *64th Annual Meeting of the American Society of Tropical Medicine and Hygiene, Philadelphia, PA, 25–29 October 2015*; 2015.

73. Babaniyi OA, Mwaba P, Mulenga D, Monze M, Songolo P, Mazaba-Liwewe ML, Mweene-Ndumba I, Masaninga F, Chizema E, Eshetu-Shibeshi M, Malama C, Rudatsikira E, Siziya S. Risk assessment for yellow fever in western and north-western provinces of Zambia. *J Glob Infect Dis* 2015;**7**:11–7.

74. Basarab M, et al. Zika virus. *BMJ* 2016;**352**:i1049.

75. Petersen LR, et al. Zika virus. *N Engl J Med* 2016;**374**(16):1552–63.
76. Campos Rde M, et al. Prolonged detection of Zika virus RNA in urine samples during the ongoing Zika virus epidemic in Brazil. *J Clin Virol* 2016;**77**:69–70.
77. Gourinat AC, et al. Detection of Zika virus in urine. *Emerg Infect Dis* 2015;**21**(1):84–6.
78. Wiwanitkit V. Urine-based molecular diagnosis of Zika virus. *Int Urol Nephrol* 2016;**48**(12):2023.
79. Stettler K, et al. Specificity, cross-reactivity, and function of antibodies elicited by Zika virus infection. *Science* 2016;**353**(6301):823–6.
80. Roehrig JT, Hombach J, Barrett AD. Guidelines for plaque-reduction neutralization testing of human antibodies to dengue viruses. *Viral Immunol* 2008;**21**(2):123–32.
81. Duong V, Dussart P, Buchy P. Zika virus in Asia. *Int J Infect Dis* 2017;**54**:121–8.
82. Abushouk AI, Negida A, Ahmed H. An updated review of Zika virus. *J Clin Virol* 2016;**84**:53–8.

Chapter 9

Post-Zika Virus Infection Survival

Zika virus infection presents as an asymptomatic or mild, self-limiting illness which is similar to Dengue fever-like disease that presents with maculopapular rash (90%), pruritis (79%), prostrations (73%), headache (66%), arthralgias (63%), myalgias (61%), nonpurulent conjunctivitis (56%), and lower back pain (51%). The concurrent fever is usually low-grade.[1] Some patients can develop severe form of disease which requires hospitalization, and other manifestations such as neurological diseases have been reported in the postinfectious period. Data from French Polynesia documented 73 cases of Guillain-Barré Syndrome and other neurological conditions subsequent to Zika virus infection.[2]

ZIKA VIRUS INFECTION AND PREGNANCY

Although there is no evidence implying increased susceptibility or the higher degree of severity in the symptoms experienced by pregnant women compared to the nonpregnant women (Fig. 9.1), Center for Disease Control and Prevention suggests that infection with Zika virus during pregnancy predisposes the fetus to congenital microcephaly and other brain defects. Other adverse pregnancy outcomes have been reported with Zika virus infection, including miscarriage and stillbirth.[3]

Lucia de Noronha et al.[4] found that placenta of pregnant women infected with Zika virus exhibited chronic placentitis along with the presence of Zika viurs in Hofbauer cells (HBC) (human placental macrophages). The activation of macrophages can result in alteration of the villous architecture and the trophoblastic epithelium, and impairment of placental immunological barrier, thus facilitating infection with Zika virus.[4] Carvalho et al.[5] identified chronic placentitis (TORCH type—which includes toxoplasmosis, other [syphilis, varicella-zoster, parvovirus B19], Rubella, Cytomegalovirus, and

Zika Virus Disease. https://doi.org/10.1016/B978-0-12-812365-2.00010-X

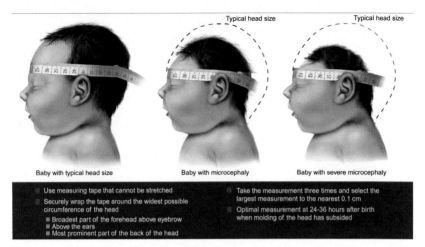

FIG. 9.1   Head circumference measurements for normal babies versus those with microcephaly. http://www.cdc.gov/zika/pdfs/microcephaly_measuring.pdf.

Herpes Simplex virus infections) with chronic villous inflammation, edema, and trophoblastic epithelial lesion. There was an increase in number of villous human placental macrophages and stromal lymphocytes and histiocytes, with calcification of the placenta. Third trimester placenta samples exhibited villous maturity, persistence of cytotrophoblasts, thickening of trophoblastic basement membrane, hypervascularity and stromal fibrosis, hyperplasia of human placental macrophages, edematous terminal villi, villous sclerosis, calcification focus with moderate increase in intervillous, and perivillous fibrinoid deposits. Chronic placentitis TORCH type that is often diagnosed during the last weeks of pregnancy appears to occur during the first trimester of Zika viurs infection. Unlike other TORCH pathogens, Zika virus infection does not cause a massive inflammatory response within the fetal and maternal circulatory system, but induces greater damage in the central nervous system of the fetus.[5]

CONGENITAL EFFECTS OF PLACENTAL ZIKA VIRUS INFECTION

The most common prenatal findings in the fetuses associated with maternal Zika virus infection were ventriculomegaly, cerebral calcifications, microcephaly, brain atrophy, brain asymmetry, and hydrancephaly.[6] Neuronal migration anomalies marked by lissencephaly, pachy/agyria, polymicrogyria, or opercular anomalies have also been reported.[6] Additionally, dysgenesis of septum pellucidum, which is associated either directly or indirectly by the ruptured septum pellucidum, were also reported.[6,7]

Ventriculomegaly—Ventriculomegaly induced by Zika virus infection most frequently demonstrated a nonhypertensive pattern, with often asymmetrical or unilateral involvement contrary to toxoplasmosis or lymphocytic choriomeningitis virus, where there is often bilateral and symmetrical involvement. In toxoplasmosis and lymphocytic choriomeningitis virus, the ventriculomegaly is persumably due to the obstruction of aqueduct of sylvius by necrotizing process. Zika virus infection-associated ventriculomegaly, is more likely related by thinning of cortical mantle is frequently observed simultaneously.[6]

The ventriculomegaly is often associated with subependymal pseudocysts around the occipital horns. Neonatal brain computed tomography scans demonstrated ventriculomegaly in almost all neonates with confirmed Zika virus infection, often involving the whole ventricular system. In 40% of the neonates, ventriculomegaly was predominantly in the lateral ventricles with enlargement of the trigones and posterior horns.[8]

Calcifications—Calcifications associated with Zika virus infection are predominantly found at the corticosubcortical white matter junction. Additional calcifications have also been identified in the midbrain, basal ganglia, brainstem, and cerebellum (similar to Cytomegalovirus and lymphocytic choriomeningitis virus). Concurrent to the calcifications, dysgenesis of cerebellum, brainstem, thalamus, basal ganglia, and spinal cord has been reported reflecting disseminated lesions induced by Zika virus infection.[6] On prenatal ultrasound, brain parenchymal atrophy is often noted along with calcifications.

Neonatal brain computed tomography scan often demonstrated intraparenchymal calcifications mainly located at the corticomedullary junction within frontal and parietal lobes. The calcifications are predominantly punctate in shape with a band-like distribution.[6,8]

Microcephaly—Following ventriculomegaly and calcifications, the third most common finding is microcephaly, which is defined as reduced head circumference (HC) < 2 and < 3 of standard deviation.[6] (Fig. 9.1) The Center for Disease Control and Prevention estimates that the rate of microcephaly following Zika virus infection ranges between 1%–13%.[3] The diagnosis of microcephaly is often made between 26 and 33 weeks of gestation.[6] The occurrence of microcephaly appears to be dependent on the timing of prenatal infection with Zika virus. Maternal infection during the first or early second trimesters are more likely to result in microcephaly than the infections at other intervals.[8] The premature closure of anterior fontanel and overlapping skull bones gives a characteristic microcephalic appearance in fetuses with Zika virus infection. The newborns tend to have excessive scalp tissue with skin folds and a normal capillary pattern probably attributed to interruption of neurogenesis and severe brain injury.[5]

Additional findings in newborns include maculopapular rash, hepatosplenomegaly, and signs of neurologic and physical disfigurement.[7]

OPHTHALMOLOGIC SEQUALAE IN FETUS, NEONATES, AND ADULTS

In utero exposure to Zika virus infection can predispose newborns to several ophthalmologic abnormalities such as chorioretinal atrophy, optic nerve abnormalities (hypoplasia and severe optic disc cupping), subluxation of lens, bilateral iris coloboma, macular atrophy, and focal pigment mottling or stippling.[7,9,10]

The most common lesion observed were focal pigment mottling and chorioretinal atrophy, especially in the macular, paramacular, and nasal retina among the newborns and infants with bilateral ocular lesions. Furthermore, two distinct characteristics are noted: (1) circumscribed areas of chorioretinal atrophy, with selective regions of choroidal vessel absence, and (2) circumscribed area of pigmentary clumping.[9] Anterior ocular manifestations are commonly found in adults (but not neonates) with acute Zika virus infection including anterior uveitis and nonpurulent conjunctivitis.

One case of hypertensive iridocyclitis related to Zika virus infection was reported in a 39-year-old male physician who presented with bilateral ocular discomfort, blurred vision, and mild erythema one week after he presented with systemic manifestations of Zika virus infection. Slit-lamp examination revealed moderate ciliary injection, mild anterior chamber reaction, and miosis of both the eyes. Intraocular pressure was high on both sides: right eye-40 mmHg and left eye-28 mmHg. He eventually experienced improvement in symptom after several weeks of treatment with topical medications (steroids, cycloplegic, and hypotensive agents).

LONG-TERM SEQUELAE

Long-term sequelae following arboviral infections are not well described. Prominent developmental, physical, and neurological defects resulting from congenital Zika virus infection are known. There may be an increase in Guillain-Barré syndrome subsequent to Zika virus infection. Other neurological manifestations include meningoencephalitis, myelitis, and facial paralysis have been reported.[11]

Guillain-Barré syndrome—Guillain-Barré syndrome is an acute paralytic neuropathy, which may be preceded by viral infection. Guillain-Barré syndrome has been described in other viral infections such Dengue and Chikungunya viral infections. The incidence of Guillain-Barré Syndrome ranges between 0.8 and 1.9% per 100,000 per year.[11] Dos Santos et al., analyzed the rate of Guillain-Barré Syndrome in Zika virus infected patients in Brazil, Columbia, the Dominican Republic, El Salvador, Honduras, Suriname, and Venezuela. They analyzed 164,237 confirmed and suspected cases of zika virus infection and 1474 Guillain-Barré Syndrome cases that occurred between April 1, 2015 and March 31, 2016.[12]

In French Polynesia, the Guillain-Barré syndrome cases occurred in association with Zika virus infection. Motor axonal neuropathy was the most common manifestation. Zika virus-associated Guillain Barré syndrome in the Pacific and parts of the United States have cranial nerve involvement similar to the Miller Fisher variant.[13] Post contrast enhancement of trigeminal and facial nerves, conus medullaris and cauda equine nerve roots, ventral roots greater than dorsal roots on brain and spine magnetic resonance imaging have been reported. The pathogenesis involves activation of both humoral and cellular immune systems. Molecular mimicry (similarity in structure) between Zika virus and myelin structures result in immune response that also effects the myelin in addition to the Zika virus. The Zika virus polyprotein resembles human proteins involved in myelination, axonal function, and neurodevelopment. Therefore, cross-reaction between Zika virus-induced neutralizing antibodies and peripheral nerve proteins such as myelin is possible.[11]

1. Bilateral T2-hypersensitivity and contrast-enhancement of the lumbar spinal ganglia.[8]

Meningoencephalitis, Encephalopathy, Acute Disseminated Encephalomyelitis.

Encephalopathy is defined as a clinical state of altered mental status or cognitive impairment with or without brain inflammation while encephalitis specifically implies underlying inflammation.[11] Few sporadic cases have been reported in the literature connecting Zika virus infection with meningoencephalitis, encephalomyelitis, and encephalopathy.[11]

A case of *acute disseminated encephalomyelitis was reported following Zika virus infection* in a 19-year-old woman who presented with gradual onset of tetraplegia, urinary retention, and reduced level of consciousness. She had suffered Zika virus infection (confirmed by polymerase chain reaction) 3 weeks prior to this presentation. Magnetic resonance imaging of the brain and spinal cord revealed multiple hyperintense fluid-attenuated inversion recovery foci in the brain and extensive longitudinal myelitis.[14]

Another 81-year-old man returning from a 4-week cruise to New Caledonia, Vanuatu, the Solomon Islands, and New Zealand developed high grade fever and coma with hemiplegia on the left side and paresis of the right upper limb. Magnetic resonance imaging revealed subcortical white matter hyperintensities on fluid attenuation inversion recovery imaging, and multiple punctuated hyperintensities on the diffusion-weighted sequences. A slight hyperintensity of the right rolandic fissure was also noted. Cerebrospinal fluid analysis following a lumbar puncture demonstrated leukocytosis suggestive of acute meningitis. Reverse transcriptase-polymerase chain reaction of the cerebrospinal fluid confirmed the presence of Zika virus infection suggesting Zika virus-associated meningoencephalitis.[15]

Two cases of encephalopathy were reported in the Caribbean islands of Martinique. The first patient was a young adult admitted with fever, arthralgia, asthenia, and headache, which were followed by seizures and a decreased level of consciousness. The second patient was an elderly patient, presenting to the hospital with acute onset mental confusion, dysarthria, right facial palsy and aphasia, conjunctivitis, bilateral hands edema, and peripheral arthritis on the initial presentation. Electroencephalogram demonstrated frontotemporal slow waves.[11]

ZIKA VIRUS INFECTION-ASSOCIATED HEMATOSPERIMA

Zika virus transmission by sexual intercourse was suggested by Foy et al.[16] He described a patient who was infected with Zika virus infection in southeastern Senegal in 2008. When the patient returned to Colorado, United States, he experienced common symptoms of Zika virus infection along with symptoms of prostitis such as blood in semen. His wife experienced symptoms of Zika virus infection days later. Considering the fact that the wife has no travel history outside of the United States, and had sexual intercourse with the husband a day after his return, transmission of the infection by semen was suggested.[16]

There is sufficient evidence to conclude that Zika virus not only causes congenital abnormalities and trigger Guilliain-Barre Syndrome, but it is also associated with ocular abnormalities, meningoencephalomyelitis, myelitis, and hematospermia. Treatment and therapeutic interventions are discussed in the following chapter.

REFERENCES

1. Amaral Calve G, dos Santos FB, Carvalho Sequeira P. Zika virus infection: epidemiology, clinical manifestations and diagnosis. *Curr Opin Infect Dis* 2016;**29**:459–66. https://doi.org/10.1097/QCO.0000000000000301.
2. Fauci AS, Morens DM. Zika virus in the Americas-yet another arbovirus threat. *N Eng J Med* 2016;**378**(7):601–4.
3. Fellner C. Zika in America: The year in review. *P T* 2016;**41**(12):778–91.
4. de Noronha L, Zanluca C, et al. Zika virus damages human placental barrier and presents marked fetal neurotropism. *Mem Inst Oswaldo Cruz* 2016;**111**:287–93. https://doi.org/10.1590/0074-02760160085. Epub 2016 Apr. 29.
5. de Carvalho NS, de Carvalho BF, Dóris B, et al. Zika virus and pregnancy: an overview. *Am J Reprod Immunol* 2017;**77**:e12616. https://doi.org/10.1111/aji.12616.
6. Vouga M, Baud D. Imaging of congenital Zika virus infection: the route to identification of prognostic factors. *Prenat Diagn* 2016;**36**:799–811. https://doi.org/10.1002/pd.4880.
7. Grossi-Soyster EN, Desiree LaBeaud A. Clinical aspects of Zika virus. *Curr Opin Pediatr* 2016;https://doi.org/10.1097/MOP.0000000000000449. 28:000-000.
8. Mehrjardi MZ, Keshavarz E, et al. Neuroimaging findings of Zika Virus Infection: a review article. *Jpn J Radiol* 2016;**34**:765–70. https://doi.org/10.1007/s11604-016-0588-5.
9. de Paula Freitas B, de Oliveira Dias JR, et al. Ocular Finding in Infants With Microcephaly Associated With Presumed Zika Virus Congenital Infection in Salvador, Brazil". *Jama Ophthalmol* 2016;**134**(5):529–35. https://doi.org/10.1001/jamaophthalmol.2016.0267.
10. Costa de Andrade G, Ventura CV, et al. Arboviruses and the eye. *De Andrade el al Int J Retin Vitr* 2017;**3**(4):https://doi.org/10.1186/s40942-016-0057-4.
11. Muñoz LS, Barreras P, et al. Zika Virus-Associated Neurological Disease in the Adult: Guillain-Barré Syndrome, Encephalitis and Myelitis. *Semin Reprod Med* 2016;**34**:273–9. https://doi.org/10.1055/s-0036-1592066.

12. Dos Santos T, Rodriguez A, Espinal MA, et al. Zika virus and Guillain-barré Syndrome – case series from seven countries. *N Engl J Med* 2016;**375**(16):1598–601. https://doi.org/10.1056/NEJMc1609015.

13. Pastula DM, Durrant JC, et al. Zika Virus Disease for the Neurointerventionist. Neurocrit Care https://doi.org/10.1007/s12028=016-0333-z.

14. Niemeyer B, Niemeyer R, et al. Acute Disseminated Encephalomyelitis Following Zika Virus Infection. *Eur Neurol* 2017;**77**:45–6. https://doi.org/10.1159/000453396.

15. Carteaux G, Maquart M, et al. Zika virus associated Meningoencephalitis. *N Eng J Med* 2016;**374**:https://doi.org/10.1056/NEJMc1602964.

16. Foy BD, Kobbylinski KC, et al. Probable non-vector-borne transmission of Zika Virus, Colorado, USA. *Emerg Infect Dis* 2011;**17**:880–2. https://doi.org/10.3201/eid1705.101939.

FURTHER READING

1. John S. Murray, PhD, MSGH, RN, CPNP-PC, CS, FAAN. "Understanding Zika Virus". https://doi.org10.1111/jspn.12164.

2. Froeschl G, Huber K, Von Sonnenburg F, et al. Long-term Kinetics of Zika Virus RNA and antibodies in body fluids of a vasectomized traveller returning from Martinique: a case report. *BMC Infect Dis* 2017;**17**:55. https://doi.org/10.1186/s12879-016-2123-9.

3. Fontes BM. Zika virus-related hypertensive iridocyclitis. *Arq Bras Oftalmol* 2016;**79**(1):63. https://doi.org/10.5935/0004-2749.20160020.

Chapter 10

Zika Virus Infection: Therapeutics

Chapter Outline

When asked "has the capacity to fight mosquitoes and conduct surveillance diminished in the United States?" Center for Disease Control and Prevention Director Dr. Thomas Frieden responded: "Imagine that you're standing by and you see someone drowning, and you have the ability to stop them from drowning, but you can't. Now multiply that by 1000 or 100,000. That's what it feels like to know how to change the course of an epidemic and not be able to do it."[1] During the 2015 Zika virus disease outbreak, there were about 2 million people reported to be infected all over the world.

VIRUSES

Friedrich Loeffler and Paul Froschin in 1898 found evidence that etiology of Foot-and-Mouth disease in livestock was in infectious particles that are smaller than any bacteria. That was the first clue that there was a genetic entity that lie somewhere in between living and nonliving states. In 1931, the two German engineers, Ernst Ruska and Max Knoll used electron microscopy that enabled them to take images of viruses. A few year later in 1935, an American biochemist and virologist, Wendell Stanley, examined the Tobacco Mosaic virus and found it to be mostly made from protein. Sometime later, this virus was separated into protein and ribonucleic acid parts. Tobacco Mosaic virus was the first virus to be crystallized and whose structure could therefore be studied in detail.[2]

Viruses have no fossil record, but it is quite possible that they have left some traces in the history of life. It has been assumed that viruses may be responsible

Zika Virus Disease. https://doi.org/10.1016/B978-0-12-812365-2.00011-1

125

for some of the extinctions seen in species identified by fossil records.[3] It was once thought that the viral outbreaks might have been responsible for mass extinctions, such as the extinction of the dinosaurs and other forms of life. It seems very unlikely, since a given virus typically causes disease only in one species or group of related species, but this hypothesis is hard to test. Even a hypothetical virus that could infect and kill all dinosaurs 65 million years ago could not have infected the ammonites (mollusc) or foraminifera (amoeboid protists) that also went extinct during the same era.

EVOLUTION OF ANTIVIRAL AGENTS

Idoxuridine (IDU) and trifluridine (TFT) are the first drugs from which the era of antiviral drug therapy began. In 1959, Prusoff[3a] described iodo-deoxyuridine synthesized by many viruses as a potential anticancer agent. In 1961, Herrmann showed that viruses could work against Herpes Simplex virus and Vaccinia virus. Kaufman created a topical treatment to relieve the symptoms of Herpetic keratitis. Two years later, Kaufman and Heidelberger[3b] also used trifluridine for the topical treatment of Herpetic keratitis. IDU and TFT are still been used in the topical treatment of Herpetic eye infections.

In the 1960s, Lee synthesized arabinosyladenine (ara-A) as a potential anticancer agent.[3c] Privat de Garilhe and de Rudder were the first investigators to show arabinosyladenine activity against Herpes Simplex virus and Vaccinia virus. Schabel[3d] described it further as an antiviral agent before it became the first antiviral drug to be used systemically, i.e., by Whitley in 1976, in the therapy of Varicella-Zoster virus infections. Adenine arabinoside is not used any longer for a number of reasons: it has low aqueous solubility, it is rapidly deaminated (removal of an amino group from a molecule) to the inactive arabinosylhypoxanthine (ara-Hx), but the primary reason is that it was superseded from the 1980s by acyclovir.[4]

As previously mentioned[4] in the early 1970s, Sidwell[4a] described ribavirin (virazole) as a broad-spectrum antiviral agent. For 30 years, researchers sought a disease against which ribavirin could be useful, until they finally found its unique application; together with pegylated interferon the medication could be used in the treatment of chronic Hepatitis C virus infections. The combination of pegylated interferon-α with ribavirin has since been used as the standard treatment for Hepatitis C virus infections, but is likely to be first complemented by and then replaced by direct-acting antiviral agents. Discovery of acyclovir was of crucial importance in the treatment of Herpes virus infections. Acylcovir was identified by Elion and Schaeffer as the first truly specific antiviral agent. Now, 35 years later, acyclovir is still being considered as the "gold standard" for the treatment of Herpes Simplex virus and Varicella Zoster virus infections.

Antiviral agents that started with the nucleoside analogs adenine arabinoside, trifluridine, adenine arabinoside, ribavirin, and acyclovir further evolved and thrived. In 1990s, the era of the acyclic nucleoside phosphonates a broad-spectrum antiviral agent started with the birth of (S)-HPMPA [(S)-9-(3-hydroxy-2-phosphonylmethoxypropyl]. The list has now grown to a large family of marketed drugs, including cidofovir, adefovir, and tenofovir, and various other prodrugs.

MECHANISM OF ACTION OF ANTIVIRAL AGENTS

Compounds that are used in the treatment of viral infections target human immunodeficiency virus, Hepatitis B virus, Hepatitis C virus, influenza virus, Herpes Simplex virus, and other herpesviruses such as Varicella Zoster virus and Human cytomegalovirus. A number of drugs act on steps that lead to the formation of the viral capsid or virion (i.e., assembly, budding, and maturation) while others, whose target is the assembled capsid or the virion, interfere with processes that affect the viral entry and uncoating. Most of the approved drugs block intracellular events affecting the synthesis and dynamics of viral proteins and nucleic acids. Within this group, viral polymerases constitute the major target for many antiviral drugs. Compounds that inhibit the replication of Herpes Simplex virus, Varicella Zoster virus, and Human cytomegalovirus include prodrugs (e.g., valacyclovir, valganciclovir, and famciclovir) that need to be phosphorylated in order to become substrates of the viral DNA polymerase. Viral enzymes (e.g., thymidine kinases in Herpes Simplex virus, Varicella Zoster virus, and a protein kinase human Cytomegalovirus) are responsible for the transformation of acyclovir, ganciclovir, and penciclovir into their monophosphate derivatives. Further phosphorylation steps are carried out by host cell kinases. The triphosphate derivatives of acyclovir, ganciclovir, and penciclovir that mimics the natural substrates of the viral DNA polymerase and are incorporated into the growing DNA chain and often terminate viral replication because they lack a 30 –OH in their ribose ring. Cidofovir is a phosphonate-containing acyclic cytosine analog that, unlike the inhibitors described here, does not depend on viral enzyme 9 for its conversion to the triphosphorylated form that competes with the dNTP substrates. Foscarnet is an analog of pyrophosphate, a product of the nucleotide incorporation reaction, and therefore behaves as an inhibitor of DNA polymerization. Unfortunately, our understanding of the mechanisms involved in resistance to acyclovir and other related inhibitors is limited by the absence of crystal structures of Herpes virus DNA polymerases.

DEVELOPMENT OF ANTIVIRAL AGENTS AGAINST ZIKA VIRUS

There is no proved treatment available for the Flaviviruses.[5] Over the past few decades, significant effort has been made to discover a drug to kill or neutralize the Dengue virus. Due to the similarities between the Zika and the Dengue viruses, we can use some of the knowledge we attained in Dengue virus related studies for drug discovery around the Zika virus. It is not unreasonable to hypothesize that drugs against these viruses could be developed. However we should be careful because the biology of the two viruses could be entirely different. Once a drug becomes available, the compound could be used as prophylaxis for travelers, family members of a household with infected individual(s), and the general population at risk during an epidemic.

First, understanding a disease's biology is very important for the development of treatment against the causative agent. For example, it is crucial to know where Zika virus replicates in patients, thereby guiding drug discovery in terms of how and where the drug should be administered during treatment. The duration of virus presence in blood also determines the therapeutic window for a circulating medication. The Zika virus has documented evidence of vertical transmission during pregnancy. To prevent microcephaly in a Zika virus-infected pregnant woman, the drug must have pharmacological property to cross through placenta and penetrate into the brain to block viral replication in the fetal brain. The inhibitor would also need to provide a systemic exposure high enough to inhibit viral replication in all the other organs. Thus, an ideal drug would have the property to pharmacologically inhibit Zika virus in brain as well as systemic sites.

Another challenge to evaluating the effectiveness of vaccines and treatments for Zika virus infection is availability of clinical subjects or experimental animal models. Because pregnant women, unborn fetuses, and newborns are one of the major target populations for Zika virus therapy, this poses a major challenge to the scientific community because of complexities of performing research in such subjects and may further delay the evaluation subsequent to development of any treatment. There is still no experimental animal model that can contract human diseases such as Zika virus infection, and manifest an immunological response and clinical syndrome similar to what is observed in clinical settings. In a recent study by O'Connor and Osorio, Zika virus infection was evaluated in Indian rhesus macaques and showed that the viral ribonucleic acid copies were detected in blood, saliva, and urine samples by reverse transcription polymerase chain reaction (RT-PCR) for up to 10 days postsubcutaneous infection. Even after 10 days, an occasional surge in viral ribonucleic acid copies was noted.[16] The experiments were also conducted in mice. Both animals are permissive to Zika virus infection; whether it mimics human infection and develops microcephaly or Guillain-Barre syndrome has not yet been determined.[17]The best substrate for initial phases of drug discovery appears to be cell lines infected with the Zika virus. The best way to develop these assays is by using high content screen, which has been used to develop inhibitors for Dengue virus in the past. It is an innovative image base assay in which small molecules against the virus are used to identify those compounds, which possess antiviral activity.[17a] This technique has been described in detail by Pascoalino et al.[17b] in which they used human cell line Huh7 (which is a human liver cell used in labs for research) in approximately 384 plates infected with the Zika virus. At the end to detect infection, indirect immunofluorescence was used with the help of a monoclonal antibody to detect E proteins of Flaviviruses.[17b] The investigators also used Alpha interferon as a reference compound because it has shown to have antiviral activity. To screen drugs, they used approximately 725 compounds from the Food and Drug Administration–approved drug library as well as using Zika-virus-infected Huh7 cells and interferon alpha 2a as a reference drug and 1% dimethylsulfoxide-treated infected cells as negative controls. Out of all the tested drugs, five agents were selected for further testing based on their selectivity index (SI), maximum activity, and effective concentration (EC50). These drugs included anticancer agents, hypolipidemics, antineoplastic, and antiemetic agents.[17a,b]

EVALUATION OF EXISTING ANTIVIRAL AGENTS

Because we are decades away from an effective vaccine against Zika virus infection, the development of treatment to limit Zika virus disease is of prime importance. To expedite the treatment plan for Zika virus infection, the National Institutes of Health is funding research to develop drug effectiveness-screening programs for existing antiviral drugs for other Flaviviruses, such as Dengue virus, West Nile virus, Yellow fever virus, and Japanese encephalitis virus. This research could examine the efficacy of existing drug compounds for potential antiviral activity against Zika virus. National Institutes of Health is also evaluating antivirals medication with activity against Hepatitis C, which is not a flavivirus, but is closely related to Zika virus. The goal is to develop a broad-spectrum antiviral drug that could be used to treat a variety of Flaviviruses, including Zika virus infection.

Chloroquine (a 4-aminoquinoline) is a weak base that is rapidly imported into acidic vesicles, increasing their pH.[6] The Food and Drug Administration approved the medication for the treatment of malaria and it has long been prescribed prophylactically to pregnant women at risk of exposure to Plasmodium (genus of parasitic protists which cause malaria). Chloroquine, which inhibits viral replication through its pH-dependent steps, restricts Human Immunodeficiency virus, Influenza virus, and Dengue virus, Japanese encephalitis, and West Nile fever infection. The antiviral effects of chloroquine were investigated on both Asian (using a Brazilian isolate) and African Zika virus infections in different cell types. In some studies, it showed a decrease in viral activity. It was observed that chloroquine treatment decreased the number of Zika virus-infected cells.[6]

In recent studies,[7] more than 774 compounds have been tested; 20 out of 774 showed decreased Zika virus infection in vitro screening assay. The inhibition of Zika virus infection in some of human cell lines, and primary human cells like cervical and placental cell lines was studied for some compounds. Some of the established antiflaviviral drugs (e.g., bortezomib and mycophenolic acid) and others drugs like daptomycin has activity against Zika virus infection. Several of these drugs reduced Zika virus infection across many cell types.[7] Food and Drug Administration-approved drugs can inhibit Zika virus infection in several human cells like genitourinary and neural origin. Data suggest that these drugs be carefully considered for trials among Zika-virus-infected patients. It is important to consider their safety profiles, and many of these drugs have been used during pregnancy for other indications, both in the United States and all over the world. Some are Food and Drug Administration category B drugs, meaning that "animal reproduction studies have failed to demonstrate any risk to the fetus and there are no adequate and well-controlled studies in pregnant women" or "Animal studies have shown an adverse effect, but adequate and well-controlled studies in pregnant women have failed to demonstrate a risk to the fetus in any trimester." Even those that are category C or D (risk not ruled out or positive evidence of risk, respectively) can be used in pregnancy when potential benefit outweighs the risk, which is likely in the case of Zika virus infection. For example, sertraline is one of the better-studied and most used antidepressants in pregnancy even though it is category C. We should also keep in mind that the use of these medications to treat Zika virus infection during pregnancy may require shorter courses and may involve gestational age windows that are different, and may have a better safety profile, than what is used for the currently accepted indications. More importantly, many of these drugs have been shown to cross the placenta (e.g., mefloquine, sertraline), allowing the opportunity to treat not only the mother but also the fetus. Furthermore, where data suggest lack of negative drug-drug interactions, clinical studies could test combinations of two or more of these to achieve maximal efficacy. It is critically important to note that many of the drugs shown to have anti-Zika virus activity could have untoward negative effects, particularly in the background of pregnancy. Therefore, their use should only be under supervision of clinical experts, preferably under research protocols. The use of these drugs in a clinical setting will obviously rely on the best diagnostic evidence, and every effort should be made to use the most sensitive and specific tests to optimize the accuracy of a Zika virus infection diagnosis. Like any treatment in pregnancy, the risks of the treatment and overtreatment have to be weighed against the risk of no treatment, in this case a devastating neurodevelopmental adverse outcome.

Several of the drugs have been shown to have antiviral activity. For example, methiopropamine has demonstrated activity to inhibit Dengue virus and also inhibited Zika virus survival.[7] Ivermectin had previously been shown to inhibit the activity of Venezuelan Equine Encephalitis virus, and several Flaviviruses.[7a,7b] Daptomycin is a lipopeptide antibiotic that inserts into cell membranes rich in phosphatidylglycerol and exerts an effect on phosphatidylglycerol-rich late endosomal membranes, blocking viral entry. The maximum plasma concentrations (C_{max}) for high-dose daptomycin exceed 180 mg/mL, and the duration to achieve C_{max} (t_{max}) was within the first 30 min of intravenous delivery.[7c] C_{max} for methiopropamine was between 24.2 and 47.2 mg/L an hour after intravenous injection. The C_{max} reported in the literature, and t_{max} was achieved within an hour of drug delivery were higher than concentrations required to inhibit Zika virus in vitro for aforementioned drugs.[7d]

Further advances in drug development for Zika virus disease continue focusing on pharmacokinetics and pharmacodynamics of many different therapeutic agents. These therapeutic agents include interferons such as interferon-α, β, and γ, all of which have shown to have inhibitory effects on Zika virus replication. Other potential targets for therapeutic agents include: (1) Inhibition of Zika virus entry into cells with duramycin-biotin; (2) Blocking the Zika virus from binding to receptors with nanchangmycin; and (3) Inhibiting the fusion of endosomes (a critical step in viral release) with obatoclax and chloroquine, both of which inhibited Zika virus infection in microvascular endothelial cells, human neural stem cells, and mouse models.[7e] Utilizing therapeutic agents that are already approved by the Food and Drug Administration is also an alternate to expedite drug development in a cost-effective manner. Some of the medications, which have demonstrated promising results, include antimalarial drugs such as chloroquine and mefloquine. Other medications that have shown promising results by inhibition of Zika virus replication in vitro are niclosamide (an antihelminth drug), azithromycin (antibiotic), and bromocriptine (antineoplastic medication).[7f–7h]

Other genetically targeted drugs target the ribonucleic acid within the Zika virus structure inhibiting replication. One of the most promising drugs in this category is called sofosbuvir. Sofosbuvir is currently clinically used for the treatment of Hepatitis C virus but has shown to be effective in treatment of Zika-virus-infected mice and in vitro using cell lines like Huh7.[7i,7j] The use of herbal drugs has also been tested such as curcumin, which is a food additive, and andrographispaniculata, which is a semisynthetic Chinese formulated herb, both showing activity against Zika virus with the former being tested in a human subject.[7e]

The additional focus is on developing and identifying therapeutic agents against Zika virus infection, which are safe from perspective of pregnancy. Prophylactic medications need to be developed for persons traveling to areas with endemic Zika virus infection or those who have a high chance of being exposed to Zika virus infection through other modes of exposure. There are also ongoing efforts on developing more effective monoclonal antibodies against Zika virus infection.

BIOMEDICAL RESPONSE AGAINST ZIKA VIRUS INFECTION

The Director of the U.S. National Institute of Allergy and Infectious Disease, Dr. Anthony Fauci stated that "Things like this tend not to go away….[Cases] may go up and down, but it's not just going to go away, so you need to start working on a vaccine now. It may be important in a year from now or six months from now, we don't know" Jan 21, 2016 in Time Health.

OX513A is a strain of the *Aedes aegypti* mosquito, created by a British company Oxitec. Its deoxyribonucleic acid contains a self-limiting synthetic gene, which produces a specific protein that alters its molecular structure, therefore killing mosquitoes in a few days. Mosquito (OX513A) has a toxin inside it which slowly accumulates in its cells. However they can be kept alive in the laboratory by administering a constant dose of the tetracycline (an antibiotic). Tetracycline suppresses the action of that synthetic gene that causes the change in its structure.[8] These adult male OX513A mosquitoes are then released into the environment. Male mosquitoes do not infect or bite humans and do not spread disease as they feed on nectar, not blood. The OX513A mosquito's mate with the female *Aedes aegypti* mosquitoes and then die within few days. The offspring of these OX513A mosquitoes, male and female, inherits the same gene and subsequently die, due to lack of natural tetracycline source. This way mosquitoes population can be controlled very effectively and specifically. As the OX513A mosquitoes are *Ae. aegypti* and only breed with *Aedes aegypti*, this results in death of released insects and their offspring.

The World Health Organization has recognized these OX513A mosquitoes as a possible tool for fighting Zika virus infection.[9] According to Food and Drug Administration, this trial, by Oxitec on OX513A mosquitoes, would unlikely be harmful on the environment. If it gets approved, the trial would release abundant male *Aedes aegypti* mosquito carrying a gene that will cause their offspring to die. In similar tests in Brazil, Panama, and the Cayman Islands,[10,11] wild populations of *Aedes aegypti* mosquitoes have been reduced by 90 percent or more within months of releasing these mosquitoes.[9–11]

OX513A, with their latest technological advances and genetic engineering, has raised opposition to genetically modified organism introduction. This is not the first United States approval of testing a genetically modified insect in pest control. Such technology was tested in the case of Pink Bollworms in the Southwest in 2009, though during that time their release largely went unnoticed outside the pest-control community.[12] But when the Food and Drug Administration released their draft opinion in March, saying that genetically modified mosquito test would probably have "no significant impact" on people or the environment, many people expressed their concerns on the public Food and Drug Administration site.[12,13]

Oxitec later explained to those queries from the people that almost all mosquitoes released would be males, which do not bite; males will die three or four days after release; their offspring are engineered to die without unnatural amounts of tetracycline. Angry public comments, however, largely dismissed these reassurances as corporate-funded research.

Some of the other people were concerned that driving down the *Aedes aegypti* population may create ecological imbalance and make room for some other menace insects to occupy the void.[14] That is a reasonable question, says Phil Lounibos of the University of Florida. He has studied relationship between *Aedes aegypti* and Asian tiger mosquito, *Aedes albopictus*. These tiger mosquitoes were once rare, but he has found that *Aedes albopictus* can take over from *Aedes Aegypti* due to these changes in ecological system. *Aedes albopictus* that mate with *Aedes aegypti* females make the females sterile and thus reduce mosquito population in other parts of Florida. Even though *Aedes aegypti* is recognized as a more potent vector for human disease, *Aedes albopictus* can also carry many of the same viruses (including Zika virus). As far as ecological concerns go, entomologist Bruce Tabashnik of the University of Arizona has no problem with eradicating *Aedes aegypti* from the Florida Keys or anywhere else in the Americas. "It's an invasive species," he says. "There are no ecological ethics violated."[14]

DENGUE VIRUS ANTIBODIES FOR ZIKA VIRUS INFECTIONS

Antibodies derived from patients who survived Dengue fever can be used to fight the Zika virus infection now and hopefully open the door for the development of a Zika virus infection treatment.[15] Those are some of the findings of researchers at the University of North Carolina's Gillings School of Global Public Health. The investigators showed that the antibodies derived from patients who survived Dengue fever and Zika virus (after neutralizing in cell cultures) can create a protective effect suggesting cross reactivity between antibodies produced after exposure.

"In essence, a therapeutic treatment using antibodies derived from selected Dengue and Zika virus survivors would protect pregnant women and others from contracting the Zika virus if they came in contact with it," said the study's principal investigator, Ralph Baric, PhD, professor of epidemiology at the University of North Carolina's Gillings School of Global Public Health.

Antibodies that are being generated from people infected with Dengue virus infection would unlikely cause any complications in patients infected with either virus. Although this antibody treatment will not be a long-term solution or produce a lifelong immunity but would be an effective short-term option to tackle the Zika virus infection during an outbreak. Thus this antibody therapy derived from survivors of another virus infection maybe a therapeutic tool and in some instances has been shown to be effective against other virus-related fetal complications. Thus these findings represent a potential treatment option for pregnant women at risk of Zika virus infection.

VACCINE DEVELOPMENT AGAINST ZIKA VIRUS INFECTION

Currently there are no proven antivirals or vaccine available for the infections. Vaccinations against different types of Flaviviruses have been shown to be effective only in reducing the symptoms of these viruses.

Recently a vaccine against Dengue virus (tetravalent live attenuated viral vaccine) successfully ended phase III trial, and made its way to several countries like Brazil, Philippines, and Mexico. Many institutes including Vaccine Research Center (VRC) has been trying to develop a vaccine against Zika virus infection. This approach is still at an early stage with plans to evaluate the Zika virus vaccine candidates in different cultures and animal models. As there is a low diversity among Zika virus strains, it may be possible to develop a single vaccine that is effective against all circulating strains of Zika virus. Other important questions are still unanswered, including the characteristics of a vaccine-elicited immune response capable of preventing infection and vertical transmission. A question remains is that whether complete immunity be required, or will a reduction in viral load will be sufficient to protect the fetus from disease. The role of assessing pre-existing immunity to other Flaviviruses needs to be addressed. The immune response to other Flavivirus infections or vaccines could obstruct or promote protective immunity of Zika virus infection vaccine by boosting the cross-reactive immunity to other Flaviviruses.

PREVENTATIVE MEASURES

In the immediate future, since there are no vaccines or proven treatments available against Zika virus infection, the best prospects for controlling Zika virus infection would be to reduce contact between the mosquitoes and humans, especially pregnant women who are at high risk. Preventative measures include:

- Eliminating or reducing the mosquito population by chemical methods like pesticides or larvicides
- Eliminating or reducing mosquito breeding sites such as standing water reservoirs
- Government penalties against standing water reservoirs
- Educating people at risk, including about the benefits of mosquito netting and topical insecticides
- Using condoms to prevent the sexual transmission of Zika virus infection in endemic areas.

Supportive management should be started on all patients with suspected Zika virus infection. This includes symptomatic treatment with appropriate hydration. Medications are used to control symptoms of fever such as aspirin and other antiinflammatory medication. If patients present with rash which can be a symptom of Zika virus infection then calamine lotion should be used to avoid itching sensations. In pregnancy, women who may have been exposed should be monitored very closely and tested regularly for any fetal abnormalities. Unfortunately, no treatment exists for children born with microcephaly except prompt recognition and supportive therapy with rehabilitation, physiotherapy, speech therapy, and occupational therapy. Breast feeding is still recommended even in areas where Zika virus infection outbreaks are occurring. As of now, transmission through breast milk has not been documented. We must be cautiously optimistic that solutions to prevent Zika virus infection and transmission will be found soon through either development of vaccine or antiviral agent or elimination of mosquitoes or combination of all these strategies.

REFERENCES

1. Herriman R. Zika virus quotes, 2016.
2. Mandal DA. *Virus history*. United Kingdom: News Medical; 2012.
3. Emiliani C. Extinction and viruses. *Bio Systems* 1993;**31**(2–3):155–9.
3a. Prusoff WH. Synthesis and biological activities of iododeoxyuridine, an analog of thymidine. *Biochim Biophys Acta* 1959;**32**(1):295–6.
3b. Kaufman HE, Heidelberger C. Therapeutic antiviral action OF 5-trifluoromethyl-2'-deoxyuridine in Herpes Simplex keratitis. *Science* 1964;**145**(3632):585–6.
3c. William W, Allen Benitez L, Goodman L, Baker BR. Potential anticancer agents. XL. synthesis of the β-anomer of 9-(d-arabinofuranosyl)-adenine. *J Am Chem Soc* 1960;**82**(10):2648–9.
3d. Schabel Jr. FM. The antiviral activity of 9-β-d-arabinofuranosyladenine (ARA-A). *Chemotherapy* 1968;**13**:321–38.
4. Ed Clercq. Milestones in the discovery of antiviral agents: nucleosides and nucleotides. *Acta Pharmacol Sin B* 2012;**2**(6):535–48.
4a. Sidwell RW, Huffman JH, Khare GP, Allen LB, Witkowski JT, Robins RK. Broad-spectrum antiviral activity of Virazole: 1-beta-D-ribofuranosyl-1,2,4-triazole-3-carboxamide. *Science* 1972;**177**(4050):705–6.
5. Lim SP, Wang QY, Noble CG, et al. Ten years of dengue drug discovery: progress and prospects. *Antiviral Res* 2013;**100**(2):500–19.
6. Delvecchio R, Higa LM, Pezzuto P, et al. Chloroquine, an endocytosis blocking agent, inhibits Zika virus infection in different cell models. *Viruses* 2016;**8**(12):1–15.
7. Barrows NJ, Campos RK, Powell ST, et al. A Screen of FDA-Approved Drugs for Inhibitors of Zika Virus Infection. *Cell Host Microbe* 2016;**20**(2):259–70.
7a. Lundberg L, Pinkham C, Baer A, Amaya M, et al. Nuclear import and export inhibitors alter capsid protein distribution in mammalian cells and reduce Venezuelan Equine Encephalitis Virus replication. *Antiviral Res* 2013;**100**(3):662–72. https://doi.org/10.1016/j.antiviral.2013.10.004.
7b. Mastrangelo E, Pezzullo M, De Burghgraeve T, Kaptein S, Pastorino B, et al. Ivermectin is a potent inhibitor of flavivirus replication specifically targeting NS3 helicase activity: new prospects for an old drug. *J Antimicrob Chemother* 2012;**67**(8):1884–94. https://doi.org/10.1093/jac/dks147.
7c. Benvenuto M, Benziger DP, et al. Pharmacokinetics and tolerability of Daptomycin at doses up to 12 milligrams per kilogram of body weight once daily in healthy volunteers. *Antimicrob Agents Chemother* 2006;**50**(10):3245–9.
7d. Nosten F, Karbwang J, White NJ, Honeymoon, Na Bangchang K, Bunnag D, Harinasuta T. Mefloquine antimalarial prophylaxis in pregnancy: dose finding and pharmacokinetic study. *Br J Clin Pharmacol* 1990;**30**(1):79–85.
7e. Munjal A, Khandia R, Dhama K, Sachan S, Karthik K, et al. Advances in developing therapies to combat zika virus: current knowledge and future perspectives. *Front Microbiol* 2017;**8**:1469. https://doi.org/10.3389/fmicb.2017.01469.
7f. Xu M, Lee EM, Wen Z, Cheng Y, Huang WK, Qian X, et al. Identification of small-molecule inhibitors of Zika virus infection and induced neural cell death via a drug repurposing screen. *Nat Med* 2016;**22**(10):1101–7. https://doi.org/10.1038/nm.4184.

7g. Retallack H, Di Lullo E, Arias C, Knopp KA, Laurie MT, et al. Zika virus cell tropism in the developing human brain and inhibition by azithromycin. *Proc Natl Acad Sci U S A* 2016;**113**(50):14408–13.

7h. Chan JF, Chik KK, Yuan S, Yip CC, Zhu Z, et al. Novel antiviral activity and mechanism of bromocriptine as a Zika virus NS2B-NS3 protease inhibitor. *Antiviral Res* 2017;**141**:29–37. https://doi.org/10.1016/j.antiviral.2017.02.002.

7i. Ferreira AC, Zaverucha-do-Valle C, Reis PA, Barbosa-Lima G, et al. Sofosbuvir protects Zika virus-infected mice from mortality, preventing short- and long-term sequelae. *Sci Rep* 2017;**7**(1):9409. https://doi.org/10.1038/s41598-017-09797-8.

7j. Mumtaz N, Jimmerson LC, Bushman LR, Kiser JJ, et al. Cell-line dependent antiviral activity of sofosbuvir against Zika virus. *Antiviral Res* 2017;**146**:161–3. https://doi.org/10.1016/j.antiviral.2017.09.004.

8. Akalib. OX513A, the GM Mosquito, a Bio-Weapon against the Zika Virus, 2016.

9. ADMINISTRATION FAD. *Oxitec Mosquito*. 2017.

10. Carvalho DO, McKemey AR, Garziera L, et al. Suppression of a field population of Aedes aegypti in Brazil by sustained release of transgenic male mosquitoes. *PLoS Negl Trop Dis* 2015;**9**(7):1–15. https://doi.org/10.1371/journal.pntd.0003864.

11. Harris AF, McKemey AR, Nimmo D, et al. Successful suppression of a field mosquito population by sustained release of engineered male mosquitoes. *Nature Biotechnol* 2012;**30**(9):828–30.

12. Milius S. FDA OKs first GM mosquito trial in U.S. but hurdles remain, 2016.

13. Authority fd. FDA Releases Final Environmental Assessment for Genetically Engineered Mosquito, 2016.

14. Gabrieli P, Smidler A, Catteruccia F. Engineering the control of mosquito-borne infectious diseases. *Genome Biol* 2014;**15**(11):535. https://doi.org/10.1186/s13059-014-0535-7.

15. Swanstrom JA, Plante JA, Plante KS, Young EF, McGowan E, Gallichotte EN, et al. Dengue virus envelope dimer epitope monoclonal antibodies isolated from dengue patients are protective against Zika Virus. *Am Soc Microbiol* 2016;**7**: 4e01123-16. https://doi.org/10.1128/mBio.01123-1619.

16. van den Pol AN, Mao G, Yang Y, Ornaghi S, Davis JN. Zika virus targeting in the developing brain. *J Neurosci: Off J Soc Neurosci* 2017;**37**(8):2161–75.

17. Fernandes NC, Nogueira JS, Ressio RA, et al. Experimental Zika virus infection induces spinal cord injury and encephalitis in newborn Swiss mice. *Exp Toxicol Pathol* 2017;**69**(2):63–71.

17a. Cruz DJ, Koishi AC, Taniguchi JB, Li X, Milan Bonotto R. High content screening of a kinase-focused library reveals compounds broadly-active against dengue viruses. *PLoS Negl Trop Dis* 2013;**7**(2):e2073https://doi.org/10.1371/journal.pntd.0002073.

17b. Pascoalino BS, Courtemanche G, Cordeiro MT, Gil LH, Freitas-Junior L. Zika antiviral chemotherapy: identification of drugs and promising starting points for drug discovery from an FDA-approved library. *F1000Res* 2016;**5**:2523.

Chapter 11

Economic Impact of Zika Virus

Chapter Outline

According to the World Bank Group, "Initial estimates of the short-term economic impact of the Zika virus epidemic for 2016 in the Latin American and the Caribbean region (LCR) are a total of US$3.5 billion, or 0.06% of Gross domestic product"[1a]. They used a few factors to calculate these numbers including significant health risks and behaviors to avoid transmission, the effects of this health crisis on economic staples like tourism, loss of worker productivity, public perceptions of risk from Zika virus including media attention and possible hysteria translating into loss of productivity, and the urgent action needed to be taken against the virus infection spread. The exact breakdown of costs by affected countries is given below:

Mexico	$744 million
Cuba	$664 million
Dominican Republic	$318 million
Brazil	$310 million
Argentina	$229 million
Jamaica	$112 million
Belize	$21 million
Other	$1.08 billion
Total	$3.48 billion

The World Bank estimated Zika virus infection will cost the world about $3.5 billion in 2016. Here is a breakdown of the costs in different countries (Fig. 11.1).

Zika Virus Disease. https://doi.org/10.1016/B978-0-12-812365-2.00012-3

	Income forgone		Fiscal revenues foregone	
	USD Mn	% of GDP	USD Mn	% of GDP
Latin America and Caribbean	3478	0.06	420	0.01
Largest impact in USD				
Mexico	744	0.06	80	0.01
Cuba	664	0.86	NA	NA
Dominican Republic	318	0.50	43	0.07
Brazil	310	0.01	75	0.00
Argentina	229	0.04	72	0.01
Significan Impacts, as % of GDP				
Belize	21	1.22	5	0.29
Cuba	664	0.86	NA	NA
Jamaica	112	0.81	27	0.19
Dominica	4	0.77	1	0.18
Dominican Republic	318	0.50	43	0.07

FIG. 11.1 World bank group (www.worldbank.org).

DIRECT EXPENDITURE

The World Bank calculated the above figures using many factors, and we will briefly analyze each one. To aid in combating Zika viru infections, World Bank provided $150 million as monetary support, which is included in the $3.5 billion.[1] The first element used to calculate the overall cost of a disease was to estimate the amount that will be utilized for prevention and treatment of disease. This amount includes payments for extra doctors, nurses, drugs, and prophylactic treatments.

Much of Zika virus infections direct costs were spent trying to control mosquitoes. In order to limit the spreading virus, the Brazilian government recruited from the army. Based on the information obtained from Brazilian media, close to 220,000 members of the armed forces accompanied by the workers from the health system were assigned to educate the people of Brazil in terms of how to eliminate mosquitoes and their breeding in and around where people live. An estimate of 3 million homes in 350 Brazilian cities were visited.[2]

In February 2016, Obama administration requested $1.8 billion from Congress to understand and fight the Zika virus infection. This funding would go toward expanding programs that control mosquitoes and research into vaccines and new public education programs especially for pregnant women. This spending is not included in the $3.5 billion since Presdient Obama's request was rejected by the House Appropriations Committee.[3]

While these direct methods are expensive, they have extra benefits, such as reducing the spread of related mosquito-borne diseases including Yellow Fever, Dengue, and Chikungunya viruses infections.

LOST PRODUCTIVITY

The loss of productivity during the Zika virus disease pandemic was an even greater concern. This is true with any other infectious disease that, when people get sick they either miss work completely or even when they are at work they are not as productive. Sick workers are not a big problem when it is only few people and there is no mass panic. However, when epidemics spread wildly, productivity in entire cities, regions, and countries takes a major toll.

In most people, Zika virus infections effects were not very severe—about one in five infected people showed signs of sickness. The World Bank assumed about four million people will be infected with Zika virus in 2016 and, since roughly 20% of the infected people get sick, the bank assumed 750,000 people in Latin America and the Caribbean would lose one week of paid work. This amount was added to the total cost of the epidemic.[3]

This number did not include the very small number of people who were infected with Zika virus and developed Guillain-Barré syndrome. The consequences of this syndrome are weak muscles and, in more severe cases, paralysis, but the syndrome is rarely fatal and spontaneous recovery occurs within 4 weeks. In the estimate by the World Bank for $3.5 billion does not include any cost incurred for Guillain-Barré syndrome cases. The reason for this omission was that, even though this disease is costly to treat, there are not many persons who are affected to contribute to financial loss.

LOSS FROM DEATH

To determine the value of a life, forensic economists use various methods. These methods are very different from what we think the value of life is. For example, forensic economists ask: "if we want to make the world safer and to prevent one more person from dying, how much would society pay?" In moral sense, life is priceless, which means that society should spend an infinite amount of money to prevent even a single death, however that is not economically possible.

There are numerous forensic studies out there that produce a range of estimates. For example, the cost of one life as calculated by the Environmental Protection Agency is $7.4 million when determining the value of clean air. Similarly, the United States Transportation Department currently estimates that the worth of saving a life is over $9 million when making decisions about the safety of United States roads and bridges.[3]

The life of a newborn is estimated a little differently. While the Zika virus infection does not appear to kill anyone, the value of life estimates is still useful because many babies born to pregnant mothers with Zika virus infection have microcephaly. This condition, in which a baby's head is much smaller than normal, results in abnormal development of the brain and many complications thereafter due to developmental retardation. Multiplying the statistical value of a life in Latin American and Caribbean countries by the number of affected babies is another economic cost of Zika virus infection and developmental retardation. Considering the fact that only a small number of babies have been infected, the World Bank did not include any estimates for death or severe impairment in its $3.5 billion estimate.

EFFECTS ON TOURISM

Fears of the Zika virus infection have a double impact in the Caribbean region, where vulnerability to and association with mosquito-borne diseases threaten an economy dependent largely on tourism. Since January 2015, when the Centers for Disease Control and Prevention first warned pregnant women to avoid travel to certain Latin American and Caribbean countries, officials have been estimating the outbreak's effect on the region's prime tourist season. According to Brazilian newspaper "O Globo," based on the reports obtained by tourist organizations in US Virgin Islands, they saw cancellations of travel upward of $250,000 in the first half of 2016.[4]

Brazil's tourism industry is suffering from the international alarm over the Zika virus infection outbreak that has infected 1.5 million people. According to reports from a Brazilian newspaper (Journal O Globe), tour operators attempted to appear optimistic that the number of tourists canceling their trips will diminish. As scientists begin to further understand the viral infection outbreak, they have received a number of "cancellation requests or travel postponements from foreign tourists, mostly pregnant women and caregivers." Tourist agencies like Blumar tell the newspaper they have seen less than 1% of their bookings cancelled, most due to the fact that a number of airlines have offered free cancellation of flights to pregnant women traveling to Zika-infected areas[5]. If the virus continues to spread throughout major Caribbean tourist destinations, it is reasonable to anticipate a continued loss to the tourism industry. One of recent cancelation of the Marlins-Pirates baseball game scheduled to take place in Puerto Rico is a good example, other examples are vacancies on Caribbean cruise ships and an interruption in Canadian and American business and professional conferences scheduled in tropical resorts during the winter of 2017.

DISEASE AVOIDANCE

As a general phenomenon, when a disease strikes, people react by avoiding affected areas. For example, when flu becomes recognized in a school, many children stay home because parents do not want their children to get affected. The Zika virus is virulent in many countries that have popular tourist destinations. Most of the $3.5 billion impact is from the expectation that tourists will avoid Caribbean and Latin American countries associated with the disease. While the World Bank believes that avoidance issues are the biggest factor in their $3.5 billion cost estimate, they caution that if tourists exhibit widespread avoidance, its estimates might go dramatically higher.

Zika virus infection outbreak cost the economy 3.5 billion in 2016, one of the largest and most expensive infectious disease outbreaks in the recent history. For example if we look at a recent outbreak of Ebola virus disease, in August 2014, a $230 million package was announced by the World Bank Group and an additional $170 million package announced in September 2014 to countries and implementing agencies, in Liberia, Sierra Leone and Guinea and were used in paying for essential supplies and drugs, personal protective equipment and infection prevention control materials, health workers training, hazard pay and death benefits to Ebola virus infection health workers and volunteers, contact tracing, vehicles, data management equipment, and door-to-door public health education outreach. These funds also are providing budgetary support to help the governments of Guinea, Liberia and Sierra Leone cope with economic impact of the outbreak, and are financing the scaling up of social safety net programs for people in the three countries.

US ECONOMIC CONCERNS

In the United States, both the civilians and authorities have realized the true impact of this disease, and concern has dramatically increased. In July of 2016, the first "Zika baby" born in the continental United States at a New Jersey hospital was diagnosed with a congenital malformation of the head.[6] The most up-to-date United States statistics, according to Centers of Disease Control and Prevention, showed 341 confirmed cases of Zika virus infection among pregnant women in the United States and its territories, and still increasing.[7] Moreover, leading researchers are growing concerned that young children infected with Zika virus are at greater risk of developing virus-related disorders than adults. According to Michael Callahan who is one of the founders of the Zika Foundation (a multinational philanthropic nonprofit organization), "We must not only protect pregnant women from mosquitoes and from sexual contact with men who may harbor the virus, but must now extend a protective shield over children for the first years of their life." The economic impact throughout the tropical Americas and Caribbean is overwhelming.[8]

Pregnant women infected by the virus have a 28% chance of delivering a child with microcephaly; these children will not develop normally and will require lifelong support, which will cause a drain on families and the medical infrastructure. The current estimated lifetime medical cost to care for a child with microcephaly is $10 million. This amount will translate to $20 billion for every 2000 children born with Zika virus-induced birth defects. Not withstanding but the effect of Zika virus infection on medical infrastructure will continue over the lifetime of these children.[1]

According to one of the reports in the Wall Street Journal, in Puerto Rico more than 1350 people tested positive for Zika virus exposure, which includes 168 pregnant women.[9] Doing simple math of counting the healthcare cost of caring for children with microcephaly, the numbers become truly detrimental for countries with both developed and developing economies. The economic consequences of the unfolding epidemic could be truly disastrous and cannot be correctly estimated because thousands of people are likely to be infected without displaying symptoms, the Journal concluded.[9]

Additionally, Zika virus infection can cause Guillain-Barré syndrome in up to 1% of those infected, and Guillain-Barré syndrome can cause muscular paralysis.[10] The cost to treat someone with Guillain-Barre Syndrome is in more than $500,000 per year, which translates to $1 billion per year for every 2000 people infected with the virus.[1] These are the closest current estimations for the cost of medical treatment only; the additional costs of diagnosis include testing pregnant women suspected to have been exposed to the virus. Furthermore, once the mother is confirmed to have the virus, there is the additional cost of ultrasound and intrapartum testing of the child.

The situation is well described in an article in the Harvard Public Health Review: "Simply put, Zika infection is more dangerous…than scientists reckoned a short time ago."[11] The president and Congress have been attempting to come to an agreement on spending $1.8 billion to fight and study the virus. Considering the active transmission of Zika virus throughout at least 62 countries and territories, including the southern United States and Puerto Rico, the reality is that the potential economic impact, not to mention human damage, that the Zika virus infection will bring is unlikely to be addressed by the $1.8 billion Congress is debating on spending.[12]

REFERENCES

1a Ulansky E. The economics of Zika; 2016, http://thehill.com/opinion/op-ed/284177-the-economics-of-zika [Accessed 06/20/16].
1. http://thehill.com/opinion/op-ed/284177-the-economics-of-zika.
2. http://www.latimes.com/world/la-fg-zika-brazil-20160213-story.html.
3. https://www.scientificamerican.com/article/how-do-we-know-the-zika-virus-will-cost-the-world-3-5-billion/.
4. https://oglobo.globo.com/brasil/virus-zika-ja-comeca-prejudicar-turismo-no-brasil-18641828.
5. http://www.breitbart.com/national-security/2016/02/11/brazils-tourism-industry-hit-hard-by-zika/.
6. http://www.nj.com/healthfit/index.ssf/2016/05/1st_baby_born_in_nj_hospital_with_zika-related_bir.html.
7. https://www.cdc.gov/mmwr/volumes/65/wr/mm6520e1.htm.
8. https://zikafoundation.org/our-team/.
9. https://www.wsj.com/articles/pregnant-with-zika-one-womans-fearful-journey-to-child-birth-1483112157.
10. https://www.cdc.gov/zika/healtheffects/gbs-qa.html.
11. http://harvardpublichealthreview.org/off-the-podium-why-rios-2016-olympic-games-must-not-proceed/.
12. https://www.washingtonpost.com/lifestyle/travel/how-zika-virus-is-affecting-caribbean-travel/2016/06/21/d4f48e20-2f1b-11e6-9b37-42985f6a265c_story.html?utm_term=.97d1f2a7de89.

Chapter 12

Psychological and Social Aspects of Zika Virus Disease

Chapter Outline

"When I told my boyfriend that the baby had problems, he said he didn't want to know anything about it," she says. She remembers him saying, "You know very well what I think about that child."

—Affected Mother[1]

Zika virus disease is usually a mild disease among adults that causes a light fever, skin rash, conjunctivitis (red eyes), muscle or joint pain, and general malaise that begins 2–7 days after the bite of an infected mosquito. Only 20% of the infected people develop symptoms of the disease. However, there is strong suspicion that Zika virus infection is linked to congenital microcephaly which is a rare condition wherein a baby has an unusually small head and will later have developmental retardation in later age. As Zika virus disease affects mostly pregnant woman, it has created a high-level of psychological stress among this population. It is quite understandable that even the thought of giving birth to a child with a malformation, regardless of whether the malformation has been caused by Zika virus disease, is certainly a stressful event for a family.

Though on one hand, the effect of disease is moderate among the adult population, the possible effect of Zika virus disease on pregnant woman and their families is expected to have a high psychological impact as they are to be considered a highly sensitive risk group. The psychological implications related to the Zika virus disease outbreak may be more dangerous than any potential acute physical effect. One of the reasons for such high level of psychological impact is the lack of medical and scientific knowledge along with inadequate information about specific treatments, all contributing to the hysteria. The psychosis generated by this kind of outbreak may have consequences in terms of healthcare setting overload and mistrust toward national authorities.

Zika Virus Disease. https://doi.org/10.1016/B978-0-12-812365-2.00013-5
143

As the saying goes "Fear of a disease, which oftentimes can prove more infectious than the disease itself, can and does have real negative economic and social effects" is quite true in this scenario.[2]

The distress and terror stemming from the possible outcomes of a disease can have disastrous effects on a country's economy and social structure. These phenomena have been observed recently in West Africa during the Ebola virus disease epidemic.

Although the Zika virus infection has national impact, the affect is disproportionately higher in low socioeconomic groups, as they live in areas with precarious sanitation conditions and irregular access to safe water, which contributes to the proliferation of mosquito-borne diseases along with limited access to health-care information and services. While pregnant, women are potentially subjected to high psychological burden. They cannot know for sure if and how Zika virus disease might affect their pregnancies and their own health. This situation is further aggravated as they do not have the right to legally terminate their pregnancies and cannot afford to have illegal but safe abortions. After delivery, they do not have the means to care for potentially affected children. The coverage of the epidemic has shown that their partners have abandoned many of these women during pregnancy or shortly after birth upon diagnosis of congenital Zika virus disease syndrome.

The groups of Brazilian residents most affected and traumatized by the Zika epidemic are low income pregnant women. They have limited access to health-care, education, and economical support from state aid agencies. This puts their health, their social protection needs, and their sexual and reproductive rights in jeopardy.[1]

ROLE OF THE GOVERNMENT DURING CRISIS

What we currently see in Brazil is a state wherein the government response may not be adequate to meet the basic health needs of entire communities and is not acting quickly enough to respond to the epidemic.

Low income pregnant women are being further cornered by conservative politicians in the Brazilian Congress. These politicians are using the Zika crisis to push their values by exacerbating the punishment for abortions. Low income pregnant women are already in the high-risk group for contracting the Zika virus infections and delivering babies with neurological singularities. They also have less access to health services. Passing the bill (PL 4.396/2016) would increase the prison term for abortions linked to microcephaly to up to 15 years. This bill limits the options given to poor pregnant Brazilian women so they are forced to carry a sick child to term and care for the baby for life. Such scenarios increase the amount of negative psychological impact the mother must bear and strains the mother financially. The Zika virus disease demonstrates the growing inequalities in Brazil between different income, gender, and geographical groups.

To assist these women cope, we need a comprehensive package of family planning and social protection for motherhood and childhood, including access to quality information regarding the epidemic and its risks; the right to legal and safe abortion if desired by the woman, given the potential psychological burden of carrying a pregnancy to term during an epidemic of yet unknown effects. Additionally for women who give birth to babies with disabilities, social protection for the woman and child, including an immediate financial assistance is required.

The constant suffering by pregnant women affected by the Zika virus infection needs to be acknowledged and addressed by the Brazilian government and public officials (Fig. 12.1).[3]

With the effects of Zika virus infection increasing, there is a real danger that more mothers will be abandoned by their partners as the number of babies born with microcephaly multiplies. In turn, women are increasingly being left to bear the tremendous financial and emotional costs associated with raising children with special needs. "Raising a child with severe disabilities is hard. Your whole life stops; you can't work", says Vera Lucia Giacometti, a psychologist who has been working at a state facility for children and young adults with severe disabilities for 16 years.[4] She further stated that the number may rise dramatically in the months ahead. Partners often abandon mothers of children with disabilities when the child is one or two years old, as the true strain of caring for a growing child with special needs starts to reveal itself.[4]

Despite such significant psychological stress and financial burden, it is remarkable to see lot of mothers willing to take up the responsibility of their children born with disabilities. One of many such great examples is Ms. Fabiane Lopes, the affected mother who gave birth to a child with microcephaly, who says fiercely she will never, ever give up her baby. No matter how hard the road ahead. "I'll go until the end, looking after her. Until the very end. My whole life will be dedicated to her," she says. "I just hope I will have health and patience. She's just starting her life. There is a long way to go".[1]

FIG. 12.1 The maternal Zika virus infection crisis has highlighted the persistence of gender inequalities. *APPhoto/Fernando Llano*

Dr. Ungar, a researcher at Dalhousie University, feels that good amount of hard work was put in to immediately tackle the Zika virus disease threat, however, not much efforts were made to terminate the things we already know that may cause a long-lasting psychological damage to the children.[5] This emphasizes the fact that despite all the furor attending the present Zika virus disease epidemic, there has been a lot of deficiency from the governmental agencies and the health-care providers in attempting to provide vital information regarding Zika virus disease to the general population, especially affected women during their pregnancy, thus creating anxiety and fear among them and limiting development of coping mechanisms.

News stories have aroused public anxiety with regards to the Zika virus infection; especially among pregnant women, their spouses/families, along with travelers and residents of the affected regions. The government has even stepped in and placed an advisory for women to delay pregnancy for the next 2 years until the disease is under control. The fallout from the advisory may last even after it has been lifted. Couples trying to conceive may be nervous about having an unhealthy baby; especially those mothers who were bitten by a mosquito during the outbreak. Individuals who express an irrational fear of infection are more likely to become socially detached.[6]

Those residing in areas where there is a high incidence of the Zika virus infection may show traits of obsessive-compulsive disorder. These residents will unnecessarily be meticulous and repetitive when cleaning items to prevent the potential for disease. It is also likely that children who have manifestations of microcephaly will be ostracized in their communities due to the lack of health education available to these residents (Fig. 12.2).[6]

FIG. 12.2   Gisele Felix, with her son in Rio de Janerio. Ms. Felix, who is 5 months pregnant, is refusing to open her doors and windows for fear of Zika-spreading mosquitoes entering her home. *PILAR OLIVARES/REUTERS*

THE ROLE OF MEDIA

"A dodgy tweet will travel halfway around the world before the truth has got its boots on."

—Sarah Khatry (intern at Himal Southasian)[7]

On February 2, 2016, the World Health Organization (WHO) declared the Zika virus disease a global emergency. On February 3, the first case of Zika virus disease transmission in the United States was identified in Dallas, Texas. Fear of its arrival had already made headlines in South Asia. Social media networks have demonstrated themselves to be unexpectedly far reaching in rapid dissemination of information, both accurate and false. There are two primary forms of users. The first kind is 'hubs' of news providers like periodicals and political figures; and the second, personal accounts of individual users. Both types form a complex network that, despite not being designed for this purpose, far outpaces any simplified projections from mathematical modelling, when it comes to the spread of information.

The Ebola virus disease outbreak in 2014 came with a series of rumors and speculation delivered as fact. Ebola virus, erroneously, was feared to be airborne, waterborne, and foodborne transmitted infection. Confusion and hype contributed to the panic (Fig. 12.3). Social media has the propensity to instigate "mass hysteria," a term coined in the 19th century that referred to the psychological phenomenon now known as Mass Psychogenic Illness (MPI).[7]

FIG. 12.3 Chato B. Stewart. The History of Zika. MentalHealthhumor.com[8]

The media going added fuel to the fire with statements like:

A new disease with an exotic name, Zika virus, is spreading 'explosively' around the world. It may be causing babies to be born with shrunken heads and brains. No one has immunity. Experts admit significant uncertainty about how the disease spreads, what symptoms it causes, or just which parts of the population face the greatest danger.[7]

Such statements could set up perfect situations for a large-scale panic about a threat that may not be such a great threat in reality.

WHO's information site was far more measured than the media. But a public statement by WHO head Dr. Margaret Chan was much more alarming in her addresses to the public, saying:

- Last year the disease was detected on the Americas, where it is spreading explosively.
- The level of alarm is extremely high. Arrival of the virus in some cases has been associated with a steep increase in the birth of babies with abnormally small heads.
- The possible links have rapidly changed the risk profile of Zika from a mild threat to one of alarming proportions.[7]

Dr. Chan's statements, which was extensively disapproved as poor form of communication, has caused the already senseless world press into a Zika frenzy grabbing the headlines in all newspapers like the following:

- Zika virus 'spreading explosively' in Americas, W.H.O. says[9]
- Zika virus is here in New York[10]
- Zika virus may not be in South Florida yet, but it has the potential to be[11]
- Zika virus: Up to 4 million cases predicted[7,12]

This untruthful panic created by WHO and the press were not completely correct. An investigation report published in the reputed journal Nature questioned the actual number of microcephaly cases reported during the Zika virus disease epidemic.

Dr. Anne Schuchat, principal deputy director of the federal Centers for Disease Control and Prevention stated that, "It is not known how common microcephaly has become in Brazil's outbreak. About three million babies are born in Brazil each year. Normally, about 150 cases of microcephaly are reported, and Brazil says it is investigating nearly 4000 cases".[7] This study also found only 270 of the Brazilian cases have been confirmed as microcephaly, and a tenth of the reported cases have been discounted as false diagnoses.[13] The basic facts about Zika virus disease suggests that even if the worst-case scenario is real, the statistical risk, even where conditions are favorable for the spread of the disease, is probably minimal. The very fact of this uncertainty is just one of the several psychological characteristics that make the threat of Zika virus disease feel much more worrisome than the evidence alone suggests.

Media headlines that use trigger words that are meant to grab the reader's attention correspondingly tug at the reader's psyche. The reader is constantly bombarded by these deceiving headlines that then alter their perception about how dangerous the Zika virus infection really is. Even if the respective story irons out the misleading headline, the reader has already linked the Zika virus infection to trigger words such as: "spreading explosively" and "four million possible victims". This makes it harder for the reader to control their fears.[14]

A major concern for the public includes not having readily accessible scientifically sound articles that clearly explain the nature of the Zika virus disease. Instead, the public social media pages and news outlets are flooded with articles written by journalist trying to profit from the Zika virus disease epidemic. These biased articles further increase the anxiety and fear felt within the communities where the Zika virus disease is prominent.[6]

Though Zika virus disease as a global emergency should be addressed in part by raising public awareness regarding prevention and treatment, in step with furthering of scientific research about the virus, this responsibility lies not only on the media, government and healthcare officials, but also on individuals as they should first verify and then share news calmly. This will aid in the rational handling of this global crisis, and the ones to come.[7]

Vector biologist Laura Harrington and chair of the Department of Entomology at Cornell University believes the current state of overreaction by the public to Zika virus disease crisis in U.S. could lead to more psychological harm along with damage to the environment due to unnecessary mass spraying of insecticides.

He further goes on to say, *"I'm very concerned that people are overreacting to the threat of Zika virus in the U.S. It is very disappointing to see the maps and information that the CDC has distributed showing unrealistic range distributions for both vectors - Aedes aegypti and Ae. albopictus. This is really doing a disservice to the public and vector surveillance programs"*.

He also raised the concern that the panic created by the media could divert the attention of the public from other serious health issues in the United States.[15]

RESPONSIBILITY OF HEALTH-CARE WORKERS

WHO has provided interim guidance for health-care providers with regards to the psychological support for pregnant women affected by Zika virus infection. Healthcare providers need to be more empathetic when providing care to pregnant patients with the Zika virus infection. The healthcare providers can accomplish this by providing basic psychosocial support to those patients who are afraid of having babies with microcephaly and also by addressing any concerns the infected pregnant mother may have. This will help the mothers accurately prepare for childbirth and help them understand what to expect post pregnancy.[16]

To boost resilience and preparedness and to reduce the impact of Zika virus disease outbreak, this set of measures should be considered. On one side, the physical spread of the disease must be contained through an aggressive control of *Aedes* species and health facilities that can provide a prompt and accurate diagnosis of Zika virus disease in all the critical areas. In addition, the psychosocial consequences of the likely outbreaks have to be confined by applying active emergency communication and support tactics.[17]

SOCIAL IMPLICATIONS OF ZIKA VIRUS DISEASE

United States responses to the Zika virus disease are fundamentally flawed. Headlines like "The War on Zika in Miami Turns To The Air" and "Genetically Modified Mosquitoes Newest Weapon in War on Zika,"[18] may be intended to signal the seriousness of the response. Zika virus disease's spread highlights the reality that the world's poorest families disproportionately bear the burden of the outbreak because they live with crowded living conditions, substandard sewage systems, and reliance on public water pumps that are often are surrounded by pools of standing water (and mosquitoes). Announcing war on the disease will project the image of sufferer being accountable for causing this disease thereby putting way too much blame in the hands of the sufferer who has the least power in these situations.[18]

With the recent campaigns conducted by the public health officials and advertisements hoarding across the city requesting the women in Zika virus infection-affected areas to avoid pregnancy, there is a unavoidable situation faced by the women who are subjected to stranger rape and intimate partner violence, may push many of them into unwanted pregnancies, in the wake of Zika virus disease, are at a higher risk of carrying the burden of additional stigma that they were unable to follow even simple public health recommendations. Furthermore, they are at further risk of exposure to new layers of horror to a sexual landscape after the recent reports highlighting the persistence of Zika virus disease in the male sperm for at least a period of six months in an infected man.[16] Another additional cause for concern, is the public behavior and reaction to individuals affected with sexually transmitted diseases like human immunodeficiency virus-acquired immunodeficiency syndrome making them an alienated or stigmatized group within the society. In countries like Brazil, where abortion is considered illegal and with recent front page headlines highlighting the pictures of babies with microcephaly affected with Zika virus disease, is actually leaving women with no other option but to choose for illegal route to end the pregnancy due to the constant terror of carrying pregnancy to full term in the light of this Zika virus infection crisis.[18]

The commencement of psychogenic illnesses has frequently been described subsequent to outbreaks of the virus infections that is closely related to Zika virus disease such as West Nile Virus in the United States. Furthermore, when the threat is signified by an emerging disease, the absence of substantial medical and scientific information along with lack of data regarding specific treatments might exaggerate the fear tenfold. The hysteria provoked by this type of outbreaks might have ramifications in the form of burden to healthcare and suspicion towards governmental officials.[17]

It is a universal fact that long term stress on mothers will have detrimental effects for their children. To reduce the psychological and social impact currently being faced by the pregnant women during this outbreak, it is essential to understand the irreversible effect that such great amounts of stress can cause on babies.

There are many interesting studies that have examined this transmission of stress from mother to child. Studies like CANDLE[18a], led by Dr. Francis Tylavsky at the University of Tennessee and Adverse Childhood Experiences (ACE study) have proved that when the expectant mothers are exposed to stress it will modify the genetic expression of the genome of the expected child thereby affecting the overall psychological development and mental well-being of the child not only in utero but also after birth. These research findings emphasize the fact that it's about time to establish the importance of child safety not only during the period of in utero but also after their birth (especially during the early susceptible years).

Along with this, it is also equally important to acknowledge and address the domestic violence towards pregnant mothers with the same aggressiveness we show towards addressing the fear of getting exposed to mosquito bites.[5]

CONCLUSION

It is very essential to dispel the misinformation and rumors surrounding Zika virus disease to reduce the stress and anxiety. For this reason, Substance Abuse and Mental Health Services Administration (SAMHSA) has outlined a tip sheet for coping with stress and anxiety related to Zika virus disease. Simple tips are:

- Staying informed by getting information from reliable sources such as healthcare provider for accurate healthcare information
- Turning to knowledgeable, trustworthy sources such as state or local health department
- Taking time away from the news to focus on things in life that are going well and that are under control to reduce stress and anxiety
- Knowing the symptoms of Zika virus disease, and speaking to healthcare provider immediately if there is a chance of contracting the virus
- Taking action to protect self and others from possible Zika virus exposure by using insect repellants and wearing clothing that covers the body from mosquito bites.
- Getting rid of sources of standing water, which could be the breeding areas for mosquitoes.[19]

During the health crisis like Zika virus disease, it is vital for public officers to share correct and appropriate data in order to ensure tranquility and trust in the minds of the public. This can be done by using simple messages and providing updates regularly; releasing information in a timely manner; addressing any rumors that may be going around; showing that expertise will help reduce anxiety and uncertainty in the community; expressing empathy to build trust and rapport in the community; and promoting stress management for pregnant women. If the affected population follows the above guidelines and organizations, media, and public health officials use responsible behavior around psychological stress, the psychological stress and social impact may be contained to a much greater extent than seen in current viral outbreaks.[19]

REFERENCES

1. Garcia-Navarro L. Zika Virus. Moms and infants are abandoned in Brazil amid surge in microcephaly. http://www.npr.org/sections/goatsandsoda/2016/02/18/467056166/moms-and-infants-are-abandoned-in-brazil-amid-surge-in-microcephalyhttp://www.npr.org/sections/goatsandsoda/2016/02/18/467056166/moms-and-infants-are-abandoned-in-brazil-amid-surge-in-microcephaly.

2. Kruskal Joshua. *Zika virus: How poverty and politics will determine its social costs*. https://intpolicydigest.org/2016/02/19/zika-virus-poverty-politics-will-determine-social-costs/; 2016.

3. Viana M, Gumieri S. Poor Brazilian women carry the greater burden of the zika virus epidemic. http://www.resurj.org/blog/poor-brazilian-women-carry-greater-burden-zika-virus-epidemic.

4. Sims S. Zika's emotional and financial burden has fallen squarely on women's shoulders. https://qz.com/682279/zikas-emotional-and-financial-burden-has-fallen-squarely-on-womens-shoulders/.

5. Ungar M. Worried about zika virus? Mother's stress also harms child. www.psychologytoday.com/blog/nuturing-resilience/201602/worried-about-zika-virus-mother-s-stress-also-harms-child.

6. Taylor-Robinson A. *Zika-association mental health burdens: Is little knowledge a dangerous thing?* https://blogs.biomedcentral.com/on-health/2016/04/20/zika-associated-mental-health-burdens-little-knowledge-dangerousthing/; 2016.

7. Khatry S. *Viral hysteria? Social media and Zika*. http://Himalmag.com/viral-hysteria-social-media-and-zika; 2016.

8. Chato B. Stewart. The history of zika. MentalHealthhumor.com https://blogs.psychcentral.com/humor/2016/11/zika-1/.

9. Stanglin D, Szabo L. WHO: Zika virus "spreading explosively" in Americas. *USA Today* 2016;. https://www.usatoday.com/story/news/2016/01/28/who-convene-emergency-meeting-zika-virus-threat/79450110/.

10. Fredericks Bob. Zika virus is here in New York. *New York Post* 2016;. http://nypost.com/2016/01/28/zika-virus-is-here-in-new-york/.

11. Teproff C. Zika virus may not be in south Florida yet, but it has the potential to be. *Miami Herald* 2016;. http://www.miamiherald.com/news/health-care/article56750508.html.

12. James Gallagher. Zika virus: Up to four million zika cases predicted. BBC News. http://www.bbc.com/news/health-35427493.

13. Declan Butler. *Zika Virus: Brazil's surge in small-headed babies questioned by report*. Nature 2016. https://doi.org/10.1038/nature.2016.19259.

14. Ropeik D. *The great Zika freak out and the psychology of fear. A new virus may be spreading more worry than disease*. *Psychology Today*. http://Psychologytoday.com/blog/how-risky-is-it-really/201601/the-great-zika-freak-out-and-the-psychology-fear; 2016.

15. Laura Harrington. Overreaction to Zika virus threat could affect psychological well-being of U.S. citizens. Cornell University, 2016. http://www.news-medical.net/news/20160518/Overreaction-to-Zika-virus-threat-could-affect-psychological-well-being-of-US-citizens.aspx.

16. http://Apps.who.int/iris/bitstream/10665/204492/1/WHO_ZIKV_MOC_16.6_eng.pdf.

17. Cenciarelli O, Carestia M, Pietropaoli S, Marco Ludovici G, Gabbarini V, Mancinelli S, et al. Zika virus: the fear travels by mosquitoes. Social and Psychological impact of the Outbreak. *Biomed Prevent* 2016;**1**:47. https://doi.org/10.19252/00000002F.

18. Sered S. *The social implications of Zika.* http://Thehill.com/blogs/pundits-blog/healthcare/291985-the-social-implications-of-zika; 2016.

18a. Völgyi E, Carroll KN, Hare ME, Ringwald-Smith K, Piyathilake C, Yoo W, Tylavsky FA. Dietary patterns in pregnancy and effects on nutrient intake in the mid-south: the conditions affecting neurocognitive development and learning in early childhood (CANDLE) study. *Nutrients* 2013;**5**(5):1511–30. https://doi.org/10.3390/nu5051511.

19 Behavioral Health Resources on Zika. Substance Abuse and Mental Health Services Administration (SAMHSA), 2016. www.samhsa.gov/dtac/zika.

Index

Note: Page numbers followed by *f* indicate figures and *t* indicate tables.